Praise for

"3Com's innovations helped create the w
story, but there's so much more to it—and this book has every up, down,
twist, and turn."
—Harry McCracken, Technology Editor, *Fast Company*, and former editor in chief of *PC World* and editor-at-large at *Time*

"*3Com* is a showcase of success and failure, of a legend that tried to be something bigger but lost its way instead. Chase and Zilber offer a look into the exhilaration and magic of innovation as well as the dark heart of Silicon Valley, where the machinations of executives make *Game of Thrones* look tame by comparison."
—Rob Enderle, the Enderle Group

"*3Com* is a must-read for anyone who wants to know how the magic of Silicon Valley works. It spans the startup, growth, and evolution of an iconic technology company many have forgotten. Most of all, it is about the exceptional executives and teams who gave the world the fundamental technology that now connects virtually everyone and everything. They are my heroes."
—Bill Krause, retired 3Com Chairman and CEO, and President, LWK Ventures

"3Com exploded when teamwork—and luck—could transform isolated computers into the networked world of today. Such magic is rarely duplicated, and it is a joy to see that magic captured in these pages."
—Howard Charney, 3Com Co-Founder, Grand Junction Founder, Cisco SVP

"In this era of brashness and arrogance, *3Com* reminds us that while technology can be the basis for a startup, it's people who make it go—or don't."
—Keith Raffel, startup CEO and bestselling novelist

"This is a must-read for Silicon Valley history buffs and anyone who wants to learn from the experiences of a team of people who lived all of the ups and downs a startup has to offer."
—Katherine Maxfield, author of *Starting up Silicon Valley*

"3Com deserves its place in Silicon Valley's pantheon. Finally, here's the book that explains why."
—Gary Morgenthaler, Partner, Morgenthaler Ventures

"*3Com* brings the reader into the lives and thoughts of the people that persevered, ultimately revealing a personal and inspiring story."
—Tam Dell'Oro, Founder and CEO, Dell'Oro Group

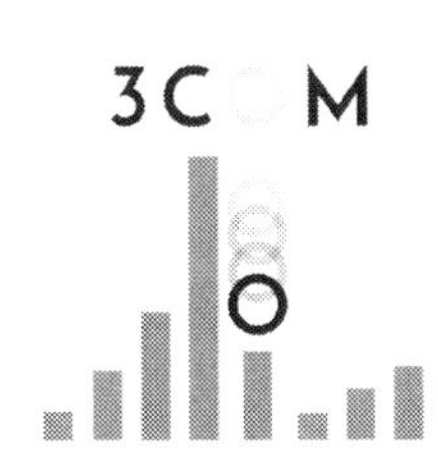
3C M
O

To Steve Sharp,
Great we both know
Michelle, a great friend of
mine - I hope you enjoy
the reading! Best wishes,
[signature]

3Com

The unsung saga of the Silicon Valley startup that helped give birth to the Internet—and then fumbled the ball

By Jeff Chase with Jon Zilber

www.3ComStory.com

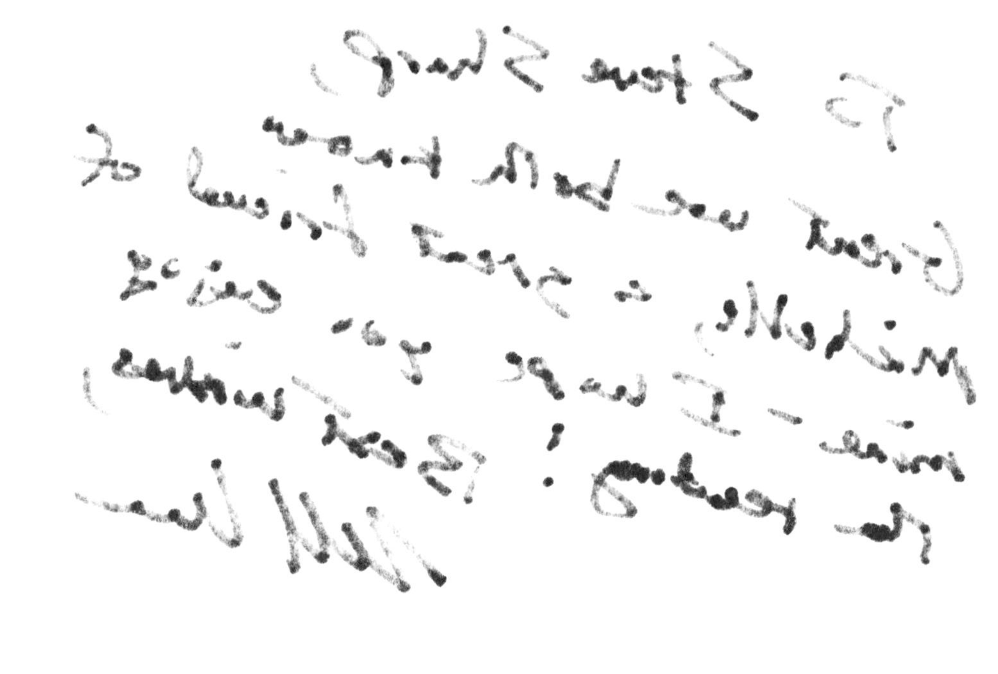

3Com: The unsung saga of the Silicon Valley startup that helped give birth to the Internet—and then fumbled the ball.

www.3ComStory.com

Hardback ISBN-13: 978-1-7330040-0-8
Paperback ISBN-13: 978-1-7330040-1-5
Ebook ISBN-13: 978-1-7330040-2-2

Cover design by Rose Zilber
Interior design by Kento Ikeda

"Don't cry that it's over.
Smile that it happened."
—Dr. Seuss

Table of Contents

Table of Figures

3Com Executive Chronology: The table below captures the leadership of 3Com over its 30 years and gives the reader a quick reference for who's who in the book. The names of people that were interviewed for this book are italicized. I also spoke with many others, who are recognized in the Epilogue and on the book website, www.3comstory.com.

Year They Joined	Notable Roles at 3Com
	Founding Fathers (and a few mothers)
1979-1980	*Robert Metcalfe, Founder, Executive, Inventor* *Greg Shaw, Co-Founder, Executive* *Bruce Borden, Co-Founder, Executive* *Howard Charney, Co-Founder, Executive* Ron Crane, Co-Founder, Executive, Inventor
1981	*Bill Krause, President and CEO* *Larry Birenbaum, VP Engineering* *Dave DePuy, Director Federal/OEM Sales* *Gerald Petak, Controller* David Spiller, VP Finance Larry Hartge, VP Product Marketing
1982	*Dan Robertson Director Manufacturing, Customer Service* John Celii, Jr, CFO Charles Kempton, VP Sales/Marketing
1983	*Debra Engel, SVP Human Resources*
1984	*Mark Michael, SVP Legal* *Cindy Hawkins, VP Corp. Controller, VP NSOps Controller* Paul Sherer, VP Technology, CTO
1985	*Chris Paisley, CFO* *Alan Kessler, SVP Sales, Service, Palm Interim CEO*
1986	*Andy Verhalen VP, GM Network Adapter Division, Director Marketing* Bob Bressler, CTO John Marman, VP Sales, VP International
1987	*Eric Benhamou, GM, CEO, President, Chairman* *Jerry Dusa, VP Sales* *Bob Finocchio, President 3Com Systems, EVP NSOps* *Edgar Masri, SVP Business Development, GM (CEO 2006)* *Judy Estrin & Bill Carrico Co-Founders, Bridge Communications* *Richard Joyce, President 3Com Europe* Steve Rizzone, VP Customer Service (via Bridge) Wes Raffel, VP Intercontinental Ops
1988	*Eileen Nelson, SVP HR* *Judy Bruner, Corporate and NSOps Controller, Treasurer (Palm SVP and CFO 1999)* *Matthew Kapp, SVP, EMEA and Asia Pacific* James Eustice, Treasurer Eugene Buechele, CTO
1989	*Ron Sege, EVP NSOps, GM (COO and Director 2008)* *Jeff Thermond, VP and GM, Network Systems Division* *Rich Redelfs, VP Mobile, WiFi* *Gwen McDonald, SVP HR* *Leslie Denend, EVP Product Ops, Corporate Development* Tom Steding, VP Enterprise Systems Division Alain Tingaud, VP 3Com Europe

	Transformers
1990	*Ralph Godfrey, SVP Sales, SVP e-Commerce* *John Hart, VP, CTO*
1992	*Doug Spreng, SVP Connectivity* *Randy Heffner, SVP Manufacturing* *Kevin Canty, Director Manufacturing* *Janice Roberts, SVP Marketing, Business Development, Palm CEO (via BICC)*
1994	*Dave Tolwinski, VP, GM, Switching Division (via Synernetics)* *Nachman Shelef, VP and GM (via NiceCom)*
1995	*Jef Graham, VP and GM Mobile, SVP Commercial and Consumer Networks* William Marr, EVP WW Sales Thomas L. Thomas, CIO
1996	Steve Rowley, VP Europe, SVP Americas Sales
1997	*Donna Dubinsky, Palm CEO* *Casey Cowell Vice-Chairman, CEO of USR, 1997* *Tom Werner, VP, Personal Connectivity* Michael Seedman, VP Personal Communications Ross Manire, SVP Carrier Systems Richard Edson, SVP Client Access *Irfan Ali, SVP Carrier Networks* John McCartney, President Client Access Mark Slaven, VP Finance, CFO 2002
1998	*Bruce Claflin, COO, CEO 2001*
1999	Dennis Connors, SVP e-Commerce John McClelland, SVP Ops David Starr, CIO Jan Soderstrom, SVP Brand Management
	Repairmen
2000	*George Everhart, SVP WW Sales* *Anik Bose, VP Corporate Business Development* Michael Rescoe, CFO Jeanne Cox, VP Marketing
2001	David Starr, CIO
2003	Shusheng Zheng, COO, H3C; Neal Goldman, SVP Legal Susan Bowman, SVP HR James Fieger, SVP WW Sales Nick Ganio, EVP WW Sales
2004	Donald Halstead, CFO Hilton Nicholson, SVP Product Ops
2005	James Hamilton, President Tipping Point Marc Willebeek-LeMair, SVP, CTO
2006	*R. Scott Murray, CEO* Robert Mao, EVP, COO, CEO Robert Dechant, SVP
2007	Jay Zager, CFO
2008	*Saar Gillai, SVP Worldwide Products and Solutions*

Foreword by Bob Metcalfe

(Internet pioneer, Ethernet inventor, 3Com founder)

The single most important new fact about the human condition is that—thanks to telegraphs, telephones, television, Internet, World Wide Web, and smart mobile phones—we are suddenly connected.

You are about to read some stories in Jeff and Jon's book about some of the people who made that happen, who helped connect more than half of the 7.6 billion people on Earth over the last 50 years. Even as we deal with the abundant connectivity that's swept over us the last 50 years, here it comes again, ready or not: the augmented video mobile gigabit Internet of Things.

I am writing this in 2019, the 50th year of the Internet. The Internet began switching packets as the ARPANET on October 29, 1969. It was my good fortune to be asked as a Harvard grad student to build a "high-speed" network interface at MIT in 1970, carrying 300 Kbps between a DEC PDP-10 minicomputer and an ARPANET packet-switching IMP. Soon our early packet protocols (e.g., TELNET, FTP) started to flow. I built another IMP interface at Xerox PARC in 1972. When PARC decided to build some of the first personal computers and put one on every desk, I was ready to write the memo inventing Ethernet on May 22, 1973. Joining me as inventors on the subsequent patent were David Boggs, Butler Lampson, and Chuck Thacker (RIP), none of whom joined 3Com when I founded it with Greg Shaw on June 4, 1979.

Let's jump back and talk for a moment about the innovation of Ethernet.

Ethernet accomplished three big things (3, that number again).

First, Ethernet took the Internet out of character-oriented minicomputer networks in false-floored computer centers into packet-oriented networks among desktop personal computers.

Second, Ethernet made bandwidth abundant, increasing the bandwidth at my Xerox desk in one day from 300 baud to 2.94 Mbps. To save you the division, that's an increase of bandwidth by not 10%, not double, not ten times, not a thousand times, but almost 10,000-fold. It was this abundance that made it reasonable years later to waste bandwidth uploading cat pictures on Facebook.

And third, Ethernet joined TCP/IP in helping drive the Internet's move to open standards. We created IEEE 802 to standardize the Ethernet local-area network (LAN). And then came the dots: 802.3, 802.4, 802.5, 802.11 (Wireless Ethernet).... Tim Berners-Lee (today's 3Com Founders Professor at MIT) came along in 1989 to add three more standards: URL, HTML, and HTTP, establishing the World Wide Web on top of TCP/IP/Ethernet.

In 1978, I attended an MIT alumni lunch at Ming's in Palo Alto to talk about raising money for startups. Shortly after, I gave seven months' notice to my boss at Xerox, David Liddle, and left to "pursue entrepreneurial ambitions." By June 4, 1979, we had formed the DIX—DEC, Intel, Xerox—alliance to make Ethernet a worldwide industry standard.

As DIX worked on the IEEE standard, I figured there was a company in there somewhere. I called my MIT fraternity brother, the only lawyer I knew, Howard Charney, to incorporate. Howard would soon join Greg and me, Ron Crane, Pitts Jarvis (RIP), and Bruce Borden, as early employees (founders?) of 3Com. (The SEC made me stop calling all of 3Com's employees "founders" after our IPO.)

Truth is I had been an Ayn Rand fan since 1959. But, having gone from Route 128 around MIT to Silicon Valley around Stanford, I was pumped to use the machinery of Free Enterprise, the startup corporation.

Corporations are too often demonized by the left, but they are fabulously successful economics technologies. And then there is profit, which is not a four-letter word. It's not, as some say, profit versus people. It's profit as proof that you are meeting human needs sustainably. I recall Bill Krause's two constants of the universe: operating profit 17%, net profit 10%, what we owed our venture investors and employee shareholders.

Innovation is a critically important and harsh teacher. 3Com's history shows many dilemmas and pivots. Early on Bill Krause and I adopted a vision of client-server computing, instead of building a Cisco-type business in networking. Bill says that was a $100 billion mistake. I say that if I were a slightly better person, there would be no Cisco.

When I "retired" from 3Com in 1990, they put a plaque on a stone outside our new HQ in Santa Clara. On it were my likeness and a quote: "The only difference between being visionary and being stubborn is whether you are right or not." It had been my position all those years that "Ethernet is the answer, what is the question?" I like calling this persistence.

Why do people risk buying products from startups offering bleeding-edge technologies? Because they hope to gain competitive advantage with them. Which leads me to what 3Com taught about the secret of progress: FOCACA. Freedom Of Choice Among Competing Alternatives. This was the regime in Silicon Valley, and 3Com prospered by more often than not serving customers better than competitors, and at a profit.

Another lesson that 3Com taught me is how hard self-awareness is. I was in the running to be 3Com CEO three times. The first time in 1979 I was the only employee and only director of the company. I chose me to be CEO. The second time was in 1982 when our Board of Directors, which I had carefully recruited and chaired, decided that Bill Krause would be a better CEO than me. The third time was in 1990 when our board chose Eric Benhamou. Left to my own devices, I would not have chosen Bill and Eric over me, but time has vindicated those two CEO appointments. I guess I can say I did a good job in building 3Com's board. Bill and Eric took 3Com to greater heights. Thanks to them for putting up with me as they did.

Circa 1982 we sold three-node Ethernet starter kits. The kits enabled three IBM-compatible PCs to share a printer, share a hard disk, and exchange emails. Customers bought the kits, and the kits worked, of course, but they were not very useful. Emails among only three PCs? I was 3Com's VP of Sales and Marketing. I had to do something to help our six-person salesforce close business. I went into a short period of deep thought and made a 35mm slide that said the systemic value of a network (V) grows as the square of the number of attached nodes (N^2). And there was a critical mass to be achieved with big enough N. So, the reason 3Com's Etherlink three-node networks were not very useful was that they were not large enough. And the remedy, we pitched, was to buy more of our products. Well, our customers believed us and grew their networks, which proved to be very useful indeed. We went public in March of 1984 based on that growth. Since 1995, that 35mm sales tool has been known as Metcalfe's Law, V~N^2. Disobey it at your peril.

Anyway, it turns out that people are what 3Com was all about. Founders. Employees and their families. Suppliers. Partners. Distributors. Customers. Shareholders. Not necessarily in that order.

Most of the fine folks who worked at 3Com did so long after I left in 1990. Thank you all. I hope it was fun.

Bob Metcalfe, Austin, Texas, January 2019

Foreword by Eric Benhamou
(3Com Chairman, President, CEO)

Young entrepreneurs today are probably not aware that 3Com existed. They simply "assume" the presence of the Internet's pervasive connectivity, much in the same way that we assume we have water and electricity in our homes. In fact, as I write this paragraph, my mouse is hovering over a 3Com mousepad with the words "The Essence of Networking," a marketing campaign that 3Com ran in the mid '90s to convey our vision of networking in the 21st century, when packets would flow naturally and effortlessly across variable pipes at precisely the optimal rate, just the same way as water flows from the Hetch Hetchy reservoir to my bathroom. Users would be totally unaware that their information travels through the air, copper, or fiber in tight little Ethernet packets whose format was defined nearly 50 years ago. The fact that 3Com is probably the technology company that contributed the most to advancing this vision and delivering on its promise is one of the greatest sources of pride that former 3Comers share when recounting their professional story.

Like all far-reaching visions, it did not happen overnight. It took an entire generation. It took the decade of the '80s for 3Com to connect the first 10 million computer users. It took the decade of the '90s to connect the next 400 million. As we rolled into the 21st century, pervasive connectivity was a tangible reality, most of it built on 3Com technology.

Young entrepreneurs today are also in search for meaning. What the history of 3Com can teach them is that the exceptionally high level of engagement 3Comers demonstrated had everything to do with them feeling they were taking part in a transformative journey, full of obstacles, desert crossings, and exhilarating moments. What made the journey fulfilling was less the wealth that we created, and much more the bond that we felt with our colleagues across the globe, the values that we shared, and the conviction we were building the most innovative networking products of our generation.

/Eric Benhamou, Palo Alto, California, January 2019

Preface

I arrived in Silicon Valley in 1983, looking for a fun wave to catch. As a young CPA, I surveyed the corporate swell and found myself in the midst of an onslaught of possibilities. Would I hang with Apple as desktop computing and PC growth was exploding, or perhaps Intel, whose hyperbolic growth seemed assured thanks to Moore's Law? Perhaps I would ride it out with Tandem Computer, with its innovations in fault-tolerant computing. Maybe even Genentech, closer to San Francisco, which was creating a brave new world of biotechnology.

The first wave I caught in 1983 was with ROLM, a leader in the changing tides of telecommunications. The skills I learned there allowed me to catch the next wave at 3Com in 1989. My nine years at 3Com were transformative for my career and personal life. And after I exited the company, it was painful to witness 3Com's descent. My colleagues and I were left with unanswered questions and the sensation of a wound brought on by those difficult years, a wound that never truly healed.

I wanted to capture and understand the 3Com story in its entirety. It felt like part of me wasn't done healing, and I was left with a scar and sense of disappointment that 3Com no longer existed. Many 3Com friends felt the same. As the company hurtled toward an exit strategy, it valiantly took on a number of challenges. Intel and Cisco were on the offensive. By 2000, the dot-com revolution was imploding. The economy was in recession in 2001, and the post-9/11 business environment was fraught with uncertainty and ambiguity. When I left 3Com in early 1999, the environment was one of enormous funding for startups by over-exuberant VCs in search of gargantuan success. I decided to chase the next wave of fiber optics telecommunications with JDS Uniphase, although that company later experienced its own implosion in 2001.

I wondered what ultimately sealed 3Com's fate—and what we could learn from it. What allowed Cisco to become a company with $50 billion in sales today, while 3Com—which was basically at parity with Cisco for many years—wound up being absorbed into the Hewlett-Packard mothership in 2009 for a relatively paltry $2.7 billion? Did we misread the networking evolution, or lack the willpower to follow our vision? Perhaps our technological innovation didn't pick the right product strategy? Marketing deficiencies? Or was it our sales channel strategy that let us down? Maybe leadership style is the secret sauce that really made the difference. Perhaps success in fast-growing markets is derived from financial strategies—fortuitous mergers, acquisitions, and the like? Or, when all is said and done, in the final analysis, is it just a crapshoot? That's what I wanted to figure out.

This will hardly be the final analysis. But, in writing it, I wanted to dive deeper than the headlines, which are all that most who remember 3Com remember. I wanted to examine the defining events from one angle and then another, curious to see how a new perspective might appear. Perhaps these events would take on new meaning with behind-the-scenes insights shared by an executive or employee. Sometimes it's interesting to ask "What if?" What if a single circumstance had been different? Would 3Com's fate have been the same, or might things have turned out completely differently?

3Com's life spanned three decades, with each decade colored by distinct challenges. 3Com's first decade was a period of creation, invention, and survival. Its second decade, which I was present for, was the most instructive in the company's history according to many people I interviewed. Vision and product decisions, along with its sales approach, were pivotal to 3Com's remarkable success, as well as its ultimate denouement. 3Com's third decade reflected the rocky times of earlier product strategies coupled with an economic recession and the bursting of the internet bubble, along with new management. This third phase culminated in the reinvention of the company as a low cost, offshore, enterprise networking company leveraging a critical partnership with the Chinese telecom giant Huawei, all of which led to 3Com finally being sold to Hewlett-Packard.

To get as many perspectives as possible and to capture the full history on paper, I reconnected and talked with old colleagues and leaders at 3Com, some of whom I've stayed in touch with over the years and others I hadn't seen for decades. I also met some new faces as I probed 3Com's three decades. I asked everyone the same three questions:

- "What occurred in your formative years that got you to where you are today?"
- "How did you arrive at 3Com?"
- Most importantly, "Tell me your story and history with the company—good, bad, or ugly."

As I'd hoped, I surfaced some new insights from this process. My memory was refreshed about a few funny moments and harrowing near-misses I'd forgotten about. But the real eye-openers were the new perspectives I found from my colleagues, and the circumstances I didn't fully appreciate at the time.

Through this process, I came to understand 3Com's place in the story of networking. Using a wedding cake as an overly simplistic metaphor, 3Com pursued the foundational layers—higher volume elements of the new network that were fast, cheap, and simple to deploy. Cisco—which emerged as the company's most significant rival—pursued the higher levels of the cake, which sold in lower volume, but were more complex, more valuable, and more profitable products.

I examined the impact of differing leadership styles—how a series of CEOs with very different personal characteristics led the company, and their downstream impacts. I also reviewed the key decisions that took us down the particular roads that shaped the company's legacy. Finally, I came to appreciate how 3Com's fate was affected by outside forces and circumstances that no corporate leader could fully control. In this sense, I found some closure.

I interviewed many of the alums starting in 2015, and continued interviewing into early 2019. Jon Zilber helped with much of the writing and editing. The book is written from my perspective, although I share the soapbox with many who hold different views. I've uncovered the flimsy nature of memories—how they distort and fade with time, and how inaccurate memories can ironically become more vivid. I've sat with our conflicting memories, digested them, and honored them in this story to the best of my fact-checking ability.

Generally, I've pieced together the thoughts of those I interviewed in chronological order, but I have given some of these colorful characters the opportunity to "hold court" and spin their stories fully. I occasionally depart from chronological order to introduce a new figure who's about to take center stage, or provide some background about the emergence of a new technology that was threatening to disrupt the previous disruptor.

When I embarked on this project, various 3Com alumni encouraged me to pursue it. The adjectives they used to describe their experiences at 3Com included words like awesome, positive, innovative, and customer-focused. They told me about the lifelong skills they learned that enabled them to go on to greater successes. So, while 3Com's fate may not have been the outcome any of us desired, there is no doubt we took away wonderful experiences, friendships, and knowledge to help us persevere in our next startups or adventures.

Many of the ups and downs along 3Com's rollercoaster ride came as a result of the specific times and places where the events unfolded, driven by the strengths, weaknesses, and idiosyncrasies of the particular cast of characters involved. The exact same story will never play out again in the exact same way. But there were many turning points where a familiar set of circumstances played out in a remarkably predictable way—with the benefit of hindsight, at least. We've highlighted a dozen of these key moments in a series of charts that may offer lessons for other companies hoping to replicate some of the 3Com magic, while avoiding some of the turbulence.

My hope is that sharing this history might help future leaders, employees, and students learn from 3Com's experience and full story. I present numerous business lessons throughout, as well as key insights in the final chapter. I would like to offer this book as a lasting legacy to the many outstanding and talented individuals who worked hard to build the company, and who were each, in one way or another, part of the journey that was 3Com.

Jeff Chase

THE NUTSHELL VERSION

Chapter 1: A Dragon Unleashed

In Silicon Valley parlance, a "dragon" is a company that performs so phenomenally well for an investing venture capital firm that the firm not only earns back its investment in that company, but earns back the entire investment fund. The return on the "dragon" investment more than covers all losses the investing firm might have made betting on other startups that did not succeed.

Out of the thousands of startups to ever start up, only a few have become dragons—so big and powerful, that they run the kingdom. Today's "unicorn" startups with $1 billion valuations might make their early employees happy and wealthy, but a dragon eats unicorns for breakfast.

3Com, created in 1979 by Bob Metcalfe, father of Ethernet, was a dragon in the networking industry. It was a tech startup back when there wasn't a whole cottage industry devoted to creating tech startups. And the impact of this dragon was far more than merely a financial one. It was founded and managed by a cast of characters who—in different times and places—might have become big-city mayors, university presidents, or beloved bartenders. Instead, this global crew (which included many first- and second-generation immigrants hailing from Algeria, Norway, England, Israel, Lebanon, Egypt, and everywhere in between) brought to life the dreams their parents had for them and, especially, their children. Their efforts unleashed the power of Ethernet which, in turn, empowered the globally connected world we take for granted today.

3Com zoomed from a rough idea to a multibillion-dollar company by commercializing what was an esoteric and radical idea at the time: a goal of connecting every computer to every other computer, to peripheral devices, and to private and later public networks. The 3Com pioneers transformed emerging technologies into off-the-shelf products. They evaluated all kinds of wild and crazy new ideas, and asked all kinds of questions. What would you use this network for, and what else could we connect it to? Why would you want to share your printer with someone else? How will we interconnect and communicate among workstations, PCs, servers, and peripherals? Just as Alexander Graham Bell's invention of the telephone—brought to life with the immortal first bit of dialog, "Mr. Watson…come here…I want to see you"—changed society, the first early interconnections between PCs ushered in a new and revolutionary universe of communications technology.

Networking tools and systems exploded the value of computers, and spawned a new breed of companies that built the infrastructure for the networked world we are now familiar with. 3Com's earliest products—network adapters—were a smash hit, providing the cash cow that allowed the company to explore other pieces of the networking puzzle over the next three decades.

But over time, network adapters became ubiquitous and basically free—just about every computer and device in today's Internet of Things has built-in networking systems using components that cost pennies instead of hundreds of dollars. 3Com had targeted the low-end, high-volume part of the networking market, which generated plenty of cash for years; although prices for adapters kept falling, the volume kept growing as the universe of devices (desktop computers! laptops! printers! servers!) that needed network connectivity kept growing.

As long as the sales volume grew faster than prices were falling, the low-end of the market was a fine target for 3Com. Until, that is, the price essentially fell to zero. Once chip manufacturers integrated basic networking functionality into every computer and peripheral (for free), 3Com's cash cow withered away.

Along the way, most of the bets 3Com made on more sophisticated products intended to compete against Cisco failed to pay off. Initially, 3Com and Cisco did not overlap—under Bill Krause's leadership, 3Com held a computing view (rather than a network-centric view) of the world, selling servers, storage, networking, and desktop gear, and was counting on what turned out to be a disastrous networking software relationship with Microsoft. Under Eric Benhamou's leadership, 3Com made attempts to catch up to Cisco, but was left in the dust, as Cisco's massive salesforce and clever marketing parlayed its incumbency from routing into switching. 3Com was left to position itself in so-called "edge" networking products, while Cisco dominated the "core"—not necessarily by 3Com's design, but by necessity. One reason for this may have been that 3Com was often contradictory in its decision-making about its rival. While it wanted to pursue the higher end of the market, like Cisco was doing, the company found success by avoiding Cisco as it pushed products for small and medium-sized businesses. If Cisco was making a bundle selling two-legged pants, 3Com would counter by creating a market for three-legged pants or one-and-a-half-legged shorts.

In its final phase, the company did manage to reinvent and reposition itself as an offshore manufacturer of low-cost products for the enterprise networking market via a critical "China Out" strategy, of which many were skeptical. But it proved to be an early case study in how U.S. technology companies could create partnerships to successfully compete in and enjoy the rewards of China's vast and fast-growing marketplace.

Despite going out with a fizzle instead of a bang, 3Com had a spectacularly good run over the three decades it existed as an independent company. From 1979 through 1993, the company led the networking race, despite the emergence of many formidable competitors, ranging from its fellow startups to established titans of the computer and telecommunications industries.

3Com was a company that modeled a well-articulated corporate culture and ethical behavior that everyone took seriously. This was a company that served as a training ground for a remarkable crew of people who went on to amazing post-3Com careers. This was a company that proved you could make mistakes and come back stronger than ever, over and over. Eric Benhamou was 3Com's CEO for much of the time I was there. When I asked him to comment on 3Com's most important contributions (below), he noted that, "Together we made these contributions to society. 3Com had a soul, unlike any other peer company in our era. Without this, you would have never wanted to write this book." We agree.

Here's a partial list of 3Com's contributions to our society as one of the early guardians of the networking galaxy:

- 3Com provided the gear for building the largest Ethernet networks the PC world had yet seen. Those networks were leveraged to create enormous economic value and competitive advantage for its customers, including Wells Fargo, Microsoft, AIG, Columbia Healthcare, and many others.
- 3Com connected more people, schools, teachers, and students with PCs than any other company in the 1990s via its award-winning network adapters, which dominated the market in the '80s and '90s.
- 3Com was an early leader in expanding the networking industry globally, bringing Silicon Valley technology to Europe, Israel, China, and elsewhere in the early 1990s. It was one of the first U.S. technology companies to build a highly successful joint venture with a partner in China, opening China markets in the 2000s.
- 3Com strategically developed more intellectual property and amassed more networking patents than any other company in the '90s, providing for a good patent offense and defense.
- 3Com created one of the first corporate venture funds dedicated to investments in networking technologies.
- 3Com created a 3Com foundation to endow academic chairs and create research centers in top universities.
- 3Com was an early leader in customized microchips of its own design for its networking products, which helped generate higher product margins and better differentiated products.

The company grew fast, stumbled occasionally, and changed course frequently trying to stay ahead of the frenetic pace of technological innovation—and its competitors. While it had a head start, it was ultimately outpaced by its archrival (and mother of all dragons), Cisco, after a squandered merger, a weighty course correction, and the impact of decisions about 3Com's early sales and product strategies.

3Com received plenty of attention from the media while it was flying high, but the full story has never been told. Its rollercoaster history offers useful insights that illustrate key business challenges still relevant to today's startups, as well as established companies seeking to stay ahead of competitors. The 3Com saga also underscores the significant impact that the style and tone of a company's leaders can have, not just on the company, but on its employees, its customers, and on the marketplace it serves. The 3Com story also reveals the impact of decisions made and decisions left unmade, as well as how external events can topple even the most successful companies.

Chapter 2: What Happened

The story of 3Com's 30-year rocking rollercoaster ride is full of twists and turns, summarized over the next few pages to familiarize you with (or refresh your memory about) what happened. There will be deeper dives into several themes—the leadership parade, pivotal decisions (about products, personnel, acquisitions, sales, and marketing), and the external forces that helped and hindered the explosive networking market.

But underlying the decisions, turning points, and pivots that punctuate the history of any company is its strategy. As the renowned Harvard professor Michael Porter has noted, strategy is "about setting yourself apart from the competition. It's not a matter of being better at what you do—it's a matter of being different at what you do."

Towards the end of 3Com's first decade of explosive growth when it seemingly could do no wrong, the company faced a dilemma as its leadership doubled down on computing and networking software. Cisco was investing in the higher-end of the networking market where massive future revenue growth would follow. Margins for 3Com's low-end products were continuing, but it was clear they wouldn't last forever. So 3Com needed a high-growth, high-margin segment of the networking market (like the arena Cisco was targeting), while continuing to differentiate themselves.

Michael Porter also aptly observed that "the essence of strategy is choosing what not to do." 3Com had initially focused on fast, cheap, and simple products, considered to be "on the edge" of the network. 3Com's network adapters, were later followed by low-cost boundary routers, stackable hubs, and switches. These products were incrementally more sophisticated and more expensive, but were still products for small and medium-sized businesses that typical customers in those businesses could easily deploy and manage. Thanks to the company's adapter dominance and healthy margins at the higher-volume edge of the network, life felt good. Cisco on the other hand, had attacked much more complex products in the center (or core) of the network, capturing the most valuable parts of the enterprise, with products that required IT departments to evaluate, install, and maintain.

As 3Com later began to emulate-Cisco-without-emulating-Cisco-too-much, the company may have misgauged how much it needed to change as it attempted to segue into new markets. These higher-end products didn't just require a more sophisticated customer with a bigger budget, but also involved a more sophisticated sales cycle. As 3Com shifted its product mix, it may have neither realized nor executed on the investment needed in sales and support, to compete head on with Cisco's massive salesforce.

3Com also needed to make decisions concerning which technologies would ultimately prevail, and allocate its R&D budget accordingly. Throughout 3Com's middle decade, FDDI, ATM,[1] and Fast Ethernet were all contenders to become "the next big thing" in networking. In addition to tectonic shifts in the core technologies, there were entirely new markets emerging. Telecommunications companies saw the import of the Internet and the merging of communications and computing. After its acquisition of US Robotics that landed the startup Palm Computing, 3Com was in a unique position to invest further in the edge of networking by expanding into the consumer home market with modems and handheld devices.

If 3Com's first decade were the "wonder years," this decade of the 1990s emerged as the "wander years," with the company pursuing new strategies in rapid-fire succession. And while the company grew at a great pace, their successes paled in comparison to Cisco's sustained growth. For a variety of reasons, neither the high-end play nor the pursuit of emerging new technologies and markets were as successful as Cisco's dominance of the networking core. The company tried to launch high-end products—often too little, too late—without investing in a more robust sales channel or leveraging the right sales skills for particular products. Nonetheless, the telecommunications industry was becoming an irresistible high-growth sales opportunity. So as Cisco moved towards telecom carrier product requirements, so did 3Com.

The biggest factor holding 3Com back seems to have been the lack of a cohesive and guiding strategy. Borrow from the Cisco playbook, or steer clear of it altogether? Target tech-savvy customers at smallish companies, or refocus sales sights on the executive suite at large enterprises? Accept lower margins

---

1. For a more detailed explanation of Fiber Distributed Data Interface (FDDI) and Asynchronous Transfer Mode (ATM), visit our website at http://www.3comstory.com/tech.

and compete in the emerging market for home networking? Or maybe "eat your children" and blow up the adapter business in favor of becoming a chip vendor, as adapters shrank to chip size?

Indeed, in its final chapter, the company finally landed on a well-articulated strategy, one that was also distinct from its competitors. This "China Out" strategy was able to bring 3Com back from a near-death experience. After the wonder years and the wander years, the company owned up to the reality that it was in its "winter years" and defined and pursued a target it could actually achieve. Via a joint venture with a Chinese company, Huawei, 3Com was finally able to make inroads into the enterprise networking market, thanks to the lower cost of its offshore-sourced products and a China green field to start with. This ultimately made it an attractive acquisition, one which the company later deemed a success.

The Leadership Parade

One factor that impacted 3Com's strategy and critical pivots was the decision-making and succession of different CEOs (and other top executives). The number of different senior leaders over the course of three decades wasn't particularly unusual. But there were stark contrasts among the CEOs—their backgrounds, experience, and, especially, leadership styles.

While nearly all CEOs demonstrate common traits (charisma, self-confidence, intellect, etc.), how they influence their followers toward effective performance can vary widely. A leader's behavior and values have a profound effect on the culture and ultimate success of a company.

- The story unfolds with Bob Metcalfe and his invention and creation of Ethernet. Bob's charismatic, assertive, and determined leadership style could be similar to Steve Jobs, an old friend of Bob's, but with fewer rough edges. His leadership characteristics played a major role in shaping the company, as will be discussed in depth.
- Bill Krause was hired by Bob Metcalfe as the first full-time CEO, overseeing a period of explosive growth. Bill's task-driven focus and management style was just what 3Com needed at the time. Bill's idea to create an annual Mission, Objectives, Strategies and Tactics Plan served as an important strategic discipline that lived on for the next thirty years of the company's existence.
- Eric Benhamou led the company during the heady 1990s when it wasn't always clear which acquired companies and technologies would be competitive and synergistic wins and which would be a waste of cash or stock. Eric reflects a mix of leadership styles, including participative, achievement,

ethical, and servant leadership theories. His introvert but collaborative engineering persona proved the right combination for the first half of the '90s. But the second half, not so much, as the book will cover. Eric influenced the ship's course in the 2000s as Chairman as well.

- Next up was Bruce Claflin, a charismatic IBM executive, hired in late 1998, who took the CEO helm of the sinking ship in 2001, dealing with an Internet bubble bursting and an urgent need to restructure the business. Bruce's skill of managing upwardly did not offset his executive team's levying of criticisms, while a focus on cash and profits trumped over product innovation. Nevertheless, he deserves credit for reinventing the company by adopting the "China Out" strategy.
- Scott Murray took over in 2006 but abruptly quit, due to concerns with 3Com's China strategy.
- Edgar Masri, an extrovert engineer and VP that had quit then returned to be the next CEO, came next. Edgar gets credit for assimilating all of Huawei 3Com (the joint venture with Huawei in China) under 3Com's umbrella, but was unable to consummate a deal with Bain Capital to take 3Com private.
- Ron Sege returned to his 3Com roots as President and COO after Edgar left, and helped find a buyer for 3Com: Hewlett Packard (HP). Bob Mao held power as the CEO to quell the concerns of its partner in China.

Pivotal Decisions

Several pivotal moments came up frequently in my conversations with members of the 3Com team. These were turning points that made the company successful during its 30-year lifespan. Some choices came and went with little fanfare and little second-guessing. For example, if you blink, you might miss a passing remark by early 3Com customer Steve Jobs that encouraged the company to use twisted-pair phone wire for Ethernet, and his later prophetic questioning of 3Com's product direction of the 1980s.

Several other pivotal moments had far more twists and moving pieces. These included:

- 3Com's important decision in early 1982 to build the first Ethernet adapter to connect IBM PCs to other networked devices and use the emerging PC reseller channel in 1983 for distribution and sales.
- The acquisition of Bridge Communications in 1987, after which a series of unforced errors gave Cisco a more than three-year lead in the router market and cemented their early products to the largest, most sophisticated enterprises.

- Innovative products and technologies developed to rally in the competition with Cisco, notably the NetBuilder II router product launched in 1992 using key RISC and ASIC chip technology and other tactics aimed at keeping 3Com the leader in cost-efficient boundary routing technology.
- A string of less well-known acquisitions helped 3Com "land and expand" outside the U.S. The $25 million 1992 acquisition of BICC Data Networks in the UK turned out to be a $1 billion hit. Synernetics on the East Coast enormously helped 3Com's switching business, although its DNA rooted in FDDI technology and its backplane design could not keep up with the Fast Ethernet threat from Cisco. 3Com also made other bets on most of the emerging technologies by developing products or acquiring companies that leveraged different networking protocols, but none of them could keep pace with the success of Cisco's Fast Ethernet switches.
- 3Com's sales approach—leveraging their third-party distributors and resellers—paid off well during the company's early years for promoting early innovative networking products. But Cisco's router incumbency advantage, its skill at solving complex network requirements, and its aggressive marketing and direct salesforce were far better suited for the largest and most-coveted enterprise customers.
- 3Com's board and CEO ignored much of their team's due diligence that raised questions about what would become of the troubled acquisition of US Robotics, resulting in financial restatements and lawsuits over what some considered to be fraudulent representations of that company's financial performance and inflated sales.
- The USR acquisition and 3Com's exit from the market for enterprise products in 2000, may be the two most pivotal turning points in 3Com's history. Exiting the largest enterprise accounts transformed 3Com from a company that prized innovation and invention into one focused on short-term profitability. This, in turn, dragged down much of the remaining business as customers dropped other product purchases.
- 3Com's decision to pivot towards outsourced R&D via a joint venture with the Chinese Telecom company Huawei fortuitously put 3Com back into the enterprise market in 2003, which helped to pique HP's interest in the company, ultimately leading to that company's 2009 acquisition of 3Com.

Outside Forces at Play

Luck, unforeseen turns, competitive maneuvers, macroeconomic trends, and other outside forces, which are beyond an individual company's control or even ability to reasonably anticipate, are part of every company's story and history. A few examples of the external circumstances that played a big role in shaping 3Com's chaotic fate include:

- Outside competitive pressures as voice and data networks began to merge (exhibited by Cisco in its purchase StrataCom in 1996 and its move into the telecommunications carrier space). This, combined with a desire for size and a doubling down on the edge of the network, helped motivate the troublesome rush to acquire USR in 1997.
- Meanwhile, as Ethernet technology evolved and started its path to commoditization, Intel became a far more aggressive competitor with substantial price cuts in 1997. That put more pressure on 3Com to find other lines of business to replace their high-volume, high-margin adapter business.
- The Internet bubble and economic recession of 2001, which made it more difficult for 3Com to find its footing, ultimately prompted a wide-scale jettisoning of people and products, leading to a cultural metamorphosis one might call "the rise of the walking dead." Restructurings, layoffs, and lawsuit payoffs were only slightly offset by the cash raised from the wildly successful IPO of Palm, a small subsidiary within USR at the time, and one of the few bright spots from the USR acquisition. But even that bright spot would have a backlash, as Palm's stock price soon crashed.

Ultimately, these factors and others contributed to the final transformation of 3Com and led to the company's sale to HP in 2009 for a small fraction of its former value. But to get to that point, follow the networking road.

THE WONDER YEARS
(1980–1989)

In today's world where everything electronic is magically connected to everything else automatically, it may be hard to appreciate the significance of Ethernet, a technology that is wholly taken for granted today. It was a major milestone that didn't merely connect computers both small and large, but also facilitated the rapid growth of countless other technologies, everything from affordable printing to the Internet.

Another aspect of the impact of Ethernet is a bit more abstract and a bit more "inside baseball." The process whereby Ethernet went from R&D to be embraced as an industry standard by multiple companies—including fierce competitors—also had an enormous impact on the way the tech industry operates today.

Ethernet was invented by Bob Metcalfe and David Boggs at Xerox PARC. While Xerox held the patents for Ethernet, Bob Metcalfe helped create an alliance among DEC, Intel, and Xerox (the DIX Alliance) that led to Ethernet becoming a public standard in September of 1980 and enabled networking among virtually all computers. For Ethernet to really take off, it would need to be ratified as an industry standard—something typically under the auspices of the IEEE, an industry association. Competitors, as well as customers, leery of adopting a proprietary standard, would be much more likely to sign onto an IEEE standard. Even the IEEE naming—a number assigned to the standard (802, in the case of Ethernet)—avoided any branding baggage that might favor one company or another. Bob explained, "The idea for an Ethernet standard was had by DEC's VP of Engineering [Gordon Bell] and me in February 1979. We then created IEEE 802 in 1980 to make the DIX Ethernet into the almost identical IEEE Ethernet. This took several years. Ethernet effectively became a standard in December 1982 with 19 companies announcing it."

As one of the first and most influential open standards, Ethernet didn't just streamline and accelerate the path to connected computer networks, it created a powerful example that demonstrated the potential of open standards and "coopetition"—an approach that accelerated technological evolution in many other realms beyond network connectivity.

3Com was founded as the company that would monetize Ethernet. Bob Metcalfe intended the name 3Com to be a shorthand for three "com" words: computer, communication, and compatibility. Initially, the company focused on networking for the nascent personal computer. Thanks to the explosion in the PC market—and the demand for network capabilities—3Com quickly became a networking powerhouse, and expanded its product line to include more sophisticated products for internetworking (the networking of networks).

Visionaries like Microsoft's Bill Gates and Apple's Steve Jobs[2] became some of 3Com's earliest customers (and advocates for the ubiquity of Ethernet). 3Com had other notable early customers including large computer manufacturers like DEC, Xerox PARC, Sun Microsystems, NCR, UNIVAC, and Hewlett-Packard. And these customers weren't just helping to drive up 3Com's bottom line—they were embracing Metcalfe's vision and incorporating Ethernet and networking capabilities into their own product lines.

But before that early and explosive success, the company first had to get off the ground. The idea of starting a company today is so commonplace that it's become a cliché, or a premise for a punchline on a TV sitcom. Cottage industries have popped up to streamline your venture—crowdfunding, pay-by-the-hour workspaces, and outsourced accounting, HR, and legal staff. You can learn how to pitch your product to VCs on *Shark Tank*. You can laugh at Silicon Valley geekiness by watching HBO's *Silicon Valley* and for better or worse, you might be inspired to find your own employees or friends crazy enough to join you in a risky proposition.

A few decades ago, however, there was a small number of players in the VC industry, and technology was driven by large institutions. It took intelligence, a strong personal network, and an awful lot of being in the right place at the right time to get a new company off the ground.

The story of 3Com starts with three pioneering entrepreneurs who saw an opportunity and seized the moment. The trio of Howard Charney, Bob Metcalfe, and a bit later CEO Bill Krause brought together totally different skills, all aimed at a common target to address an emerging market with innovative products. Technically, 3Com was founded by four people—Metcalfe, Charney, Greg Shaw, and Bruce Borden—but the real energy behind the company in its first decade was Metcalfe, Charney, and Krause (who was brought in from HP to serve as the company's first president). Ron Crane also rounded out the team via his key contributions to the IEEE and many brilliant, if sometimes late-to-the-party, engineering feats.

Bob launched 3Com on June 4, 1979. Bob had called on his MIT Sigma Nu fraternity brother Howard Charney, then a lawyer, who incorporated 3Com Corporation for $300. Howard went on to carry out a wide variety of roles for 3Com. Bob recalled:

> I tried to found 3Com on a napkin at the Bella Vista Restaurant on Skyline overlooking Silicon Valley. I invited John Shoch, Ron Crane,

---

2. Steve Jobs also pushed Apple's own proprietary networking technology called AppleTalk, building it into the motherboard. It had a long run, from 1984 until well into the 1990s.

> and Greg Shaw. John said no, in order to become Assistant to the President of Xerox. Ron Crane (RIP) said he had some work to finish up at Xerox, where we all used to work. Ron joined later as VP Hardware. Greg Shaw said yes, as VP Software, and so we two founded 3Com and got founder's common stock at incorporation.
>
> Bruce Borden joined later to work with Greg Shaw to deliver UNET, the first commercial version of TCP/IP, which we shipped for Unix in 1981 (with Microsoft's Steve Ballmer co-announcing). Other key players were Pitts Jarvis (RIP), another MIT Sigma Nu, and Larry Birenbaum, with whom I was on the MIT varsity squash team, both MIT '69.
>
> 3Com began operations as a consulting company out of my apartment at Oak Creek Apartments, opposite Stanford, in Palo Alto. We would later move to 3000 Sand Hill Road. Robyn Shotwell also lived at Oak Creek. She and I met in August 1979, married in February 1980, and are still married. She and I published a book at 3Com, the 3Com Local Area Network Vendors List (later Ethernet Handbook). Every VC who visited 3Com felt compelled to buy a copy of this book for ~ $250. 3Com's first 'product'!

And while early founders Greg Shaw and Bruce Borden didn't make the same mistake as the lesser known Apple Co-Founder Ron Wayne (who sold what would have been a $75 billion stake in Apple for a paltry $2,000), they left the limelight after exiting 3Com with their small claim stake and pursued other interests in the early 1980s. Before leaving, Shaw took a two-month sabbatical to sail around the world, then left to spend his claim stake pursuing his first true love, music, starting up a music studio in San Francisco. He returned to work and created a second fortune at Microsoft, and now divides his time between Brazil and Seattle. Borden has served as CEO, scientist, founder, and director for a vast array of technology companies.

Each member of the trio brought complementary skills to the table. Metcalfe brought his visionary, evangelistic, and energetic presence to every meeting, and could attract talent based on his reputation as the architect of Ethernet. Charney, coming from IBM, had the skills to be a Chief Operating Officer from day one—able to manage every operational function, from legal to manufacturing and everything in between. Bill Krause, a disciplined HP executive, had a knack for driving effective processes and was able to instill corporate discipline into the young venture. And if Metcalfe was 3Com's "Steve

Jobs"—the visionary and chief evangelist—Ron Crane (and later Paul Sherer) acted as the company's "Steve Wozniak," with their endless engineering ability and knack for ingenious product solutions.

As you read on, if you feel like the landscape shifts dramatically in the blink of an eye, that's because it frequently will. Imagine you're on a scenic tour, and your guide says, "The first part of the journey will be on horse-drawn carriage, then we'll get into a car, then an airplane, and the final leg of our journey will be in a submarine. If everything looks completely different when you look out the window, don't panic—it's supposed to."

In its first decade, 3Com started life as a consulting company, then became a networking adapter company for large computer companies like DEC, and finally a networking adapter and software company selling products that small businesses and eager department managers could buy at their local computer store, all wanting to share resources like printing or to try out this new thing called "email." But 3Com's real breakout product was shipped in September 1982, the Etherlink, an Ethernet adapter for the IBM PC and compatibles. It came with the EtherSeries network operating system software.

But soon, the company was aiming to emulate another pioneering tech company, Sun Microsystems (later subsumed by Oracle), using its special networking sauce to sell decked-out computer equipment that supported client-server computing to connect diskless workstations to servers with software to run email. Most importantly, 3Com could market the favorable economics of sharing office printers, large disc drives, and other new capabilities made possible by this newfangled networking.

The idea that people would pay a hefty premium for computers simply because they came with networking capabilities built in may seem bizarre today. Ethernet connectivity is an invisible part of all our lives. Packets of information zip around the globe with just a click or string of words spoken to Siri or Alexa or Cortana—from our homes, offices, cars, or airplane seats. But it wasn't that long ago when networking was something exotic. If you wanted it, you expected to pay for it, and to think carefully about your strategy to minimize the cost. Consider the heyday of the telegraph, when people *did* think twice before adding an extra few words to their messages, or your grandparents' or parents' experience of long-distance telephone calls for which the family would gather around the telephone to keep call duration to a minimum. (If your email inbox is like mine, however, you might agree that perhaps paying by the word is an idea worth resurrecting.)

It's easy to appreciate how ubiquitous and transformative Ethernet has become. It's harder to fathom a world without it, or appreciate how it came about. To fully understand Ethernet, it helps to understand Robert Melancton Metcalfe, aka Bob Metcalfe.

Chapter 3: Founding Fathers

Bob Metcalfe, Father of Ethernet

Bob Metcalfe was born into an adventurous and ambitious family with Viking roots, perhaps contributing to his own visionary and sometimes stubborn nature. His grandparents arrived in New York City around 1900, hailing from Oslo, Bergen, Leeds, and Dublin. His parents were born in Brooklyn. His father was an engineer and technician, as well as a steadfast union member. Bob also remembers his father as a competitive ice dancer in Brooklyn in the 1940s, "He was also a great roller-skating dancer on the indoor rink circuit. He was so smooth, so athletic, twirling our Mom on family ice outings." According to Bob, his parents' two goals in life were first, to retire one day, and second, to send Bob to college, making him the first member of the family to do so. Bob also noted that his grandmother fought organized crime on the docks of New York, so they were regaled with Mafia stories growing up.

Bob's enthusiasm for technology started at an early age. His father had a shop in the basement, with various tools and an old oscilloscope. He would work with his father to repair things like broken televisions. As Bob relates it, one day he reached into the back of a TV and inadvertently found the high voltage. His family later found him unconscious. (Perhaps that experience instilled in him his appreciation for the power of technology?)

In the fourth grade, he was assigned a book report. Having left it to the last minute, he wandered down to his dad's basement and found a book on electrical engineering, written by MIT professors. He finished his report with the words, "And someday, I intend to go to MIT and get a degree in electrical engineering." True to his word, Bob received his BS degrees in Electrical Engineering and Industrial Management from MIT in 1969, an MS in Applied Mathematics from Harvard in 1970, and a PhD in Computer Science (Applied Mathematics) from Harvard in 1973.[3]

In the eighth grade, Bob worked to create a type of computer that would use switches to replicate the function of an adding machine. (This was in 1959,

---

3. Shustek, Len. "Oral History of Robert Metcalfe." Computer History Museum. November 2006 and January 2007.

five years before the IBM 360 came out.) His teacher gave him an A++++ superior grade for his efforts.

While Bob was at MIT, he worked for Raytheon as a computer programmer in assembly language. The two or three dollars per hour he received seemed like a lot, enough so that he could hire a fraternity brother to do his laundry to make more efficient use of his own time. Bob liked the puzzle aspects of programming, and still enjoys math puzzles. At MIT, his major evolved from architecture to pure math, to physics, to management, and, finally, fulfilling his fourth grade promise, he landed on electrical engineering.

Bob noted, "When my Dad heard I was going to graduate MIT in Industrial Management in 1968, he said, 'We didn't send you to MIT to get a degree in management.' So, I stayed an extra year and added EE, two BS degrees. My father saw the world as labor vs. management."

Harvard Professor Joe Lassiter recalled how Bob Metcalfe, Howard Charney, Greg Shaw, and he were constantly running into each other through their fraternity days on campus, and later heard stories about what they were doing after graduation. Joe pointed out, "Bob was an incredibly charismatic, effective guy at communicating the big picture. He always had kind of big ideas and big insights. That was a characteristic of him from very early on. He was a person that people were naturally drawn to and naturally wanted to follow. Bob was always clearly an electrical engineer, and clearly interested in software. When Bob was pushing the development of Ethernet, He would keep talking it, pursuing it, and he would keep selling it. He was passionate about it and tireless in pursuing this visionary quest."

Metcalfe did well academically. After collecting his two bachelor's degrees, he applied and was advised to go to Harvard for his PhD, a school he had come to hate. His offer from Harvard included an assignment to help design the Interface Message Processor (IMP) interface to connect Harvard's PDP-10 computer to the IMP. After he arrived, however, he was frustrated to learn that Harvard now felt it was "too important" to have a student do the job, which they had now assigned to the IMP's original designers, the firm of Bolt, Beranek, and Newman. Ironically, the firm then brought on another student to do the actual work. (BBN Technologies, as it's known now, is a versatile R&D company, sometimes called the "third university of Cambridge," along with MIT and Harvard. Among its many notable projects was an acoustical analysis of the assassination of John F. Kennedy done as a part of a Congressional investigation.)

After his frustrating rejection from the project, Bob found work at MIT while he was still pursuing his PhD at Harvard. He offered to hook up MIT's IMP to their PDP-6 and, later, PDP-10 computers. The pay was better, and he felt it was a good way to show Harvard that they had made a mistake by not hiring him. He did some of this work with Bob Bressler who would later end up at 3Com.

Metcalfe was slated to receive his PhD in June of 1972. He had collected nine job offers, which were predicated on his finishing his degree. In May of 1972, he was finishing his thesis defense. He had initially written his doctoral thesis on how the ARPANET, the predecessor to Internet, worked. His initial PhD thesis was rejected by Harvard for not being theoretical enough; however, he had already accepted employment at Xerox's Palo Alto Research Center (PARC). Bob went on to say that Harvard "forced me to invent Ethernet," as will become clear shortly. Bob to this day has not forgiven the way Harvard treated engineers and especially him, fifty years later.

Ethernet began to germinate as a concept when the Department of Defense created the Advanced Research Projects Agency (ARPA) in 1958 to fund high-risk, long-term research. NASA was formed shortly after that as well. In the 1960s and '70s, many large companies were also investing in long-range research. IBM (Watson Research Center), AT&T (Bell Labs), and Xerox (PARC) were committed to innovation, regardless of whether they could project any short-term financial gains that might result from those efforts.

As outlined in Walter Isaacson's book *The Innovators: How a Group of Hackers, Geniuses and Geeks Created the Digital Revolution*, the first network that connected computers arrived in 1969. Legendary players such as Lawrence Roberts (RIP), Stephen Crocker, Vinton Cerf, Bob Taylor, Leonard Kleinrock, and others all contributed. On October 29, 1969, in the same year NASA sent a man to the moon and Silicon Valley built the microprocessor, ARPA created the first network that could connect distant computers.[4]

These research centers all made huge advances across a wide range of technologies. And despite the absence of clear profits, the return on investment was often substantial. For example, even though Xerox ultimately "gave away" Ethernet by allowing their patents to be used in the open standards for the technology, the company's profits from the development of laser printing alone paid back their entire investment in PARC, according to John Shoch, who became Xerox's Office Systems President. Xerox's generosity has no doubt been the gift that keeps on giving in personal computing. In addition to

---

4. Isaacson, Walter, "The Innovators: How a group of hackers, geniuses, and geeks created the digital revolution," 2014. Also, refer to https://bit.ly/2EYv1iu regarding the date of the first ARPANET message sent on October 29, 1969.

Ethernet and laser printing, Xerox developed the graphical user interface that played a major role in making the personal computer personal. PARC may not have intended to be so generous—many feel that Steve Jobs borrowed more than his fair share of various aspects of Xerox technology—but PARC has also benefited from the expansion of a tech-hungry economy.[5]

Xerox PARC's work for ARPA would often attract top PhD students. Many stayed on beyond their graduate research programs to continue work on these projects. As Bob Metcalfe noted in Judy Estrin's book *Closing the Innovation Gap*, "ARPA was a magical entity." In the 1970s, PARC was a magnet for technologists with open minds and acute vision. Michael Dolbec, 3Com's Vice President of Business Development, had this to say about that era:

> I remember how cool it was to be in the PARC room that was all white boards, and a few bean bag chairs. I just thought that was the coolest thing—where people would go in there, dream up a future, and architect in it. I think I was there for 20 hours a day for a while.

It was in this environment that Metcalfe engineered a hardware interface for putting PARC on the ARPANET (the network that connected ARPA computers), similar to the work he had done earlier at MIT's Laboratory for Computer Science. As Bob sardonically observed, "The packet-switching Internet (called ARPANET to start) began operating in 1969, only to be invented by Al Gore in 1991."

Metcalfe's incessant curiosity helped lead him to investigate the existing network technology of the time, which would later help with his work on Ethernet at Xerox PARC. Late in 1972, when he was jetlagged and unable to sleep while traveling as an ARPANET contractor at PARC, he was looking for something to read at the home of a friend (Steve Crocker, an ARPA program manager). He came across a paper analyzing the performance of a packet radio network called ALOHAnet, a network at the University of Hawaii. Bob took issue with some of the ideas proposed by that paper's author, Norm Abramson (a well-regarded communications engineer whose work would be recognized with a prestigious IEEE Alexander Graham Bell Medal and other awards).

Bob reworked the analysis model, ultimately writing his own paper that evolved into his Harvard PhD dissertation in May of 1973. The genius behind Abramson's model was the idea of terminals that would retransmit using randomization to ensure that data packets wouldn't collide with each other

---

5. As of 2017, the lab employs 260 physicists, computer and social scientists, and engineers, and carries a $500 million budget. "A new boss ponders the past and future of the fabled Xerox PARC." LA Times. January 2017.

on the network. Bob found a way to adjust the retransmission interval so that it was "just right"—not too slow, but not fast enough to cause network instability. There were other important differences. Notably, ALOHAnet could not transmit and receive simultaneously, and could not detect collisions as they occurred, making Ethernet much more efficient. His PhD thesis was based on these improvements, which he called "Packet Communication."[6]

In 1973, Butler Lampson, one of the founding members of PARC, introduced Metcalfe to Charles Simonyi, who was working on developing a network for Xerox. Butler asked Bob to take this networking project over, and asked Charles to instead develop an editor documentation program named "Bravo," which was the first "what you see is what you get" (WYSIWYG)[7] document preparation program, the precursor to Microsoft Word. Charles later went on to write MS Word at Microsoft, and needless to say, it all turned out well—his net worth is estimated to be $1.4 billion.[8] Meanwhile, Bob started designing a high-speed local network to connect Xerox's newly developed Alto computers and on-site printers. The PARC Ethernet sent its first packets in November of 1973, and evolved over the next two years, with a patent issued in 1975.

Some aspects of networking are quite sophisticated, while other aspects were more akin to tinkering around in a garage workshop. Having loaded up on 1,000 feet of coax cable, the medium for the network, Bob Metcalfe was experimenting in PARC's basement on different ways to connect and send digital bits down the cable. This would require the help of David Boggs, who, at the time, was a graduate student at Stanford, and knew how to put connectors on the cable, which involved stripping, clamping, crimping, and soldering, work that Bob had never done. That day in the basement, Bob and David Boggs bonded through the medium of coax. Later, to add a PC to the physical network, they'd drill a hole in the cable, and puncture it. The idea for this kind of "passive tap" (later descriptively called a "vampire tap") was suggested by David Liddle, who led development of the Xerox Star computer system; he first became familiar with coax while installing cable for

6. If you wish to read it, his thesis can still be found on Amazon—look for "Packet Communication" by Robert M. Metcalfe.

7. Before WYSIWYG software, special mark-up codes were added to indicate how text or images should appear—that is, what you'd see onscreen would be a jumble of text and commands that dictated things like what should be bolded or in italics (à la HTML). The phrase behind the acronym—"What you see is what you get"—was popularized in the 1970s by the comedian Flip Wilson. Dressed in drag, he played a woman named Geraldine who would sling those words so memorably on TV. For a more detailed explanation of Simonyi and WYSIWYG, visit our website at http://www.3comstory.com/tech. For more about Flip Wilson, you're on your own.

8. Charles has also made two trips into space, visiting the international Space Station in 2009.

TVs as a graduate student in Toledo, Ohio. Tapping cables to link computers to networks continued into the early 1990s.

In today's wireless world, it is easy to think of networking as an automatic presence—something that is always magically available, without any physical or visible hardware. WiFi and cellular connections just happen. Aside from entering the occasional password, there usually isn't even any software configuration required. But back in 1973, all these network connections required physical hardware and a lot of configuration. The costs and headaches of standalone networking products ultimately laid the foundation for the seamless, built-in, automatic, and far more powerful networks we take for granted today.

Ethernet's debt to ALOHAnet is significant, but there were many fundamental differences. For example, Abramson had used radio signals as the transmission medium for ALOHAnet, while a key aspect of the design of Ethernet was that it could be deployed far more broadly. Radio wasn't an option for Ethernet due to the size of the printed circuit card in the Alto computer. And radio wasn't a practical option to achieve the data transmission speed that would be needed.

In a memo dated May 22, 1973, Bob Metcalfe renamed the "Alto Aloha Network" to Ether Net, which he soon was writing as Ethernet. As Bob noted, "If Ethernet were invented on any one day, that would be on Ethernet's 'birthday,' May 22, 1973, in my memo of that day naming it and drafting how it might work." Though Bob considered naming the network technology CoaxNet after the cable used to connect network nodes, he wanted to give it a more abstract name because he felt that his technology could work using any transmission medium that was passive, readily available, and capable of propagating electromagnetic waves, as Bob illustrated in a sketch that would have long-lasting impact.[9]

The name Ethernet was derived from a concept taught in his freshman physics class at MIT—"luminiferous aether"—which was a hypothetical medium pervading all of space. It was once assumed that some kind of "ether" must exist everywhere since light waves couldn't possibly be propagated through empty space. That assumption, however, proved to be wrong. The celebrated Michelson-Morley experiment of 1887 demonstrated the opposite—light could indeed travel in a vacuum. Modern physics, which is based on relativity and quantum theory, no longer presumes the need for this hypothetical substance. Today, the notion of "ether" is merely a figurative concept. Bob also pointed out that "the Ethernet memo shows a 'radio ether,' which you can think of as

---

9. The complete 1973 memo written by Bob Metcalfe at Xerox can be found at 3comstory.com/tech.

From Luminiferous Aether to Ethernet to WiFi

Bob Metcalfe's 1973 vision for Ethernet

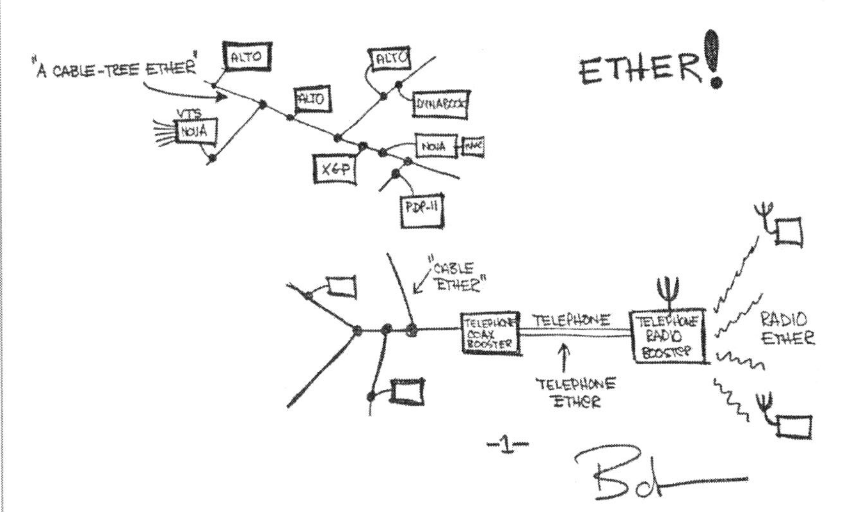

It was just a hand-drawn sketch, but the original vision for Ethernet encompassed not just the most immediate applications for networking, but long-range possibilities such as WiFi that would take decades to become practical (and ubiquitous) realities. *Image provided courtesy of PARC, a Xerox company.*

WiFi in 1973, but it took twenty-five years to get the semiconductors to make Wireless Ethernet practical. Wireless Ethernet was renamed WiFi in 1999."

Bob, along with David Boggs, Chuck Thacker, and Butler Lampson, filed for the Ethernet patent issued in 1975. In 1976, Bob and David Boggs published their seminal paper, "Ethernet: Distributed Packet Switching for Local Computer Networks."[10]

What's Wrong with Floppies?

Early 3Comers Bruce Borden and Greg Shaw recalled how the process of transforming Ethernet from theory to product was uncharted territory. It wasn't just the mechanics of making it work that was terra incognita; helping people understand the point of connecting computers was also often a foreign concept. Borden described how the two worked together, "We put the plumbing together to make that work. We had to invent all of that. In the very early days, as we were starting to demonstrate local networks, it was amazing how many people we talked to, companies that we talked to, who said, 'What's wrong with moving floppy disks between computers?' The concept of a network was just so foreign. The very first reason for adopting networks was email. A little bit of file sharing but mostly just communications. At 3Com, we had a room with a lock on it and a company that we were working with brought a machine over for us to look at putting an Ethernet into—that was Apple and the Lisa. Then IBM dropped off their PC. There wasn't ubiquitous computing yet, so the machines we were connecting together were more work stations and small cluster class machines with dumb terminals attached."

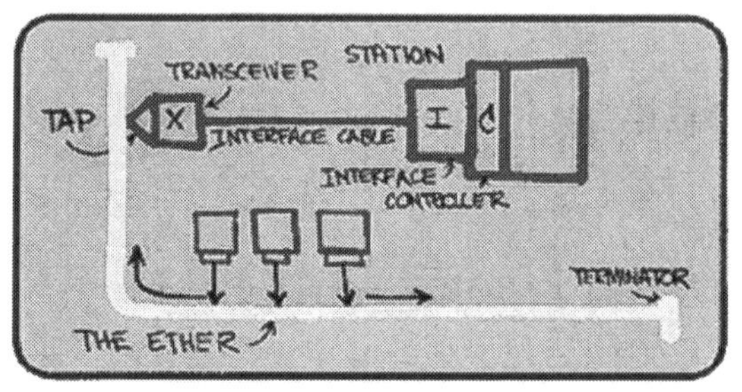

Metcalfe's sketch of Ethernet, used in his presentation to the National Computer Conference in June 1976 was formalized in their 1976 paper.

Bob Metcalfe also noted at this time, "I bought an early IBM PC and left it outside 3Com Engineering to tempt our engineers to change our product plans to include the IBM PC. Bingo!"

Greg Shaw also weighed in on those early days while working on the ILLIAC IV project[11] at the NASA Ames Research Center in Mountain View during the

10. The paper can be found on this book's companion website at 3comstory.com/tech.
11. The Illinois Automatic Computer (ILLIAC) was a series of supercomputers built at various locations, between 1951 and 1974.

summers of his freshman and sophomore years at MIT. He was introduced to Charles Simonyi, and worked with him on programming DEC computers. After Shaw graduated from MIT and Charles had moved to work at Xerox PARC, Charles invited Shaw to work with him on the Bravo word processor project. While at Xerox, after hearing Alan Kay describe his vision of the Dynabook[12] that he and Alan Kay were working on, Shaw wanted to build his own device that would compose music automatically, which he called the Dynastudio. When he explained his idea to start his own company to Metcalfe, Metcalfe responded by offering to work together, but at 3Com. Greg recalled:

> I thought he was merely joking, but a year later he called me up to say that it was time to start our company. We started out in Bob's apartment, giving regular parties for all of the Silicon Valley movers and shakers we could convince to come by and chat with us. Bob filled his bathtub with ice, wine, and beer, and the rest of the party flowed easily.
>
> Nothing concrete appeared, other than a good time, until a couple of months later when Bob got a call from General Electric, inviting us to consult for Jack Welch, GE's then President. [GE ultimately proved a waste of time].
>
> We rented an office on Sandhill Road, so as to be close to the venture capital companies that had coalesced there, and set about building a few Ethernet adapters for a small number of mini-computers on the market, as well as the software to run the network on Unix (which is where Bruce Borden comes in, with whom I spent a delightful year doing so). Soon thereafter, the IBM PC came out. Recognizing the bombshell it represented, we moved all the focus on to enabling Ethernet for the PC, and the rest is history.

How a Bob Becomes a Law

If the genius idea for bootstrapping the company from an idea into a commercial reality was evangelizing Ethernet as an industry standard, the recipe for ratcheting up sales volume stemmed from another Metcalfe vision—the one with his name on it.

---

12. This Dynabook could be considered the role model for laptops, tablets, and smartphones. In January of 2007, Jobs announced essentially this type of tablet device, but then shrunk it to handheld size, with a phone packed in it. It ran on OS X, which had originally come from NeXT. For a great read on this history, check out Fire in The Valley, by Michael Swaine and Paul Freiberger.

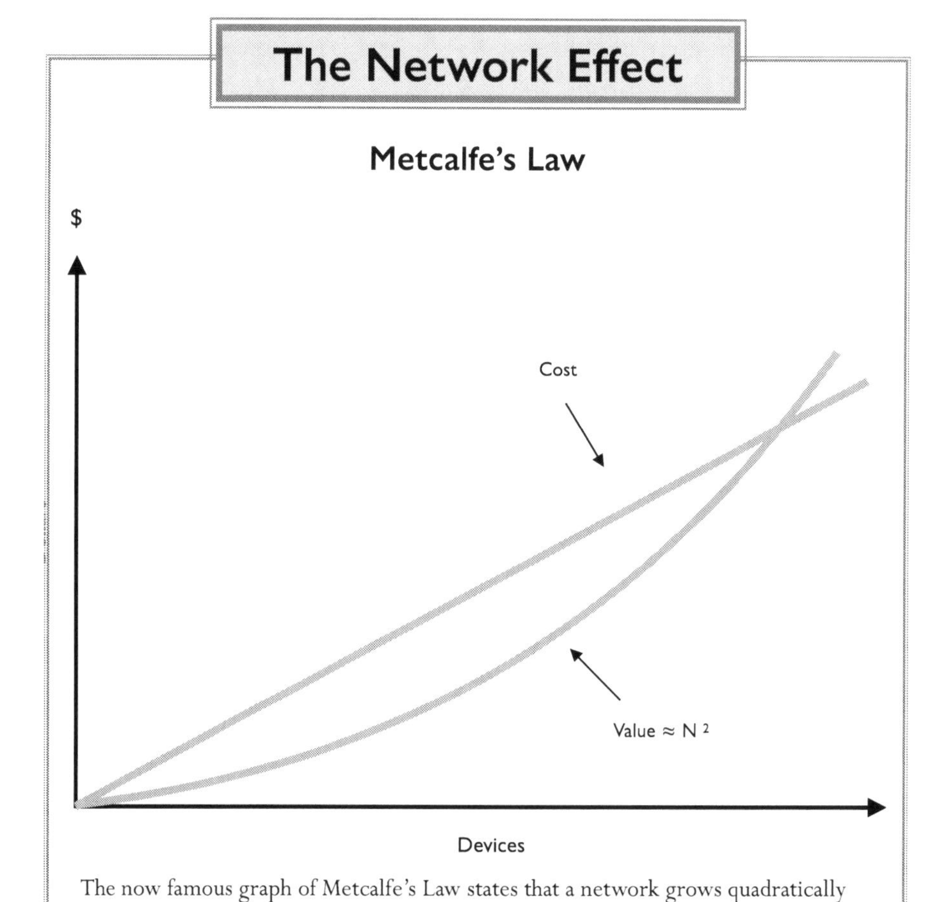

The now famous graph of Metcalfe's Law states that a network grows quadratically more valuable as people or devices are connected. As the size of the network grows from say, 50 to 100 people, the value of the network grows by much more than double.

Today, Bob Metcalfe is perhaps best known for his formulation of Metcalfe's Law, which states that the value of a network increases as the square of the number of users connected to the system. A network with 10 people has the value of 10^2 or 100. A network with 100 people has the power of 100^2 or 10,000.

And as we've seen with the emergence of billion-user networks like Facebook, their value exceeds all expectations (except, perhaps, those of Bob Metcalfe). The power of such networks rivals that of many nations. Metcalfe's Law, not yet so named, began life as a 35mm slide, a sales tool for selling Ethernet to PC buyers in the early 1980s. It was named Metcalfe's Law in 1993 by one of the leading technology gurus of the day, George Gilder, and written about in his influential 1995 book *Telecosm.*[13]

And if you're in the business of selling devices for connecting machines to networks, it pays to convince customers of the value of Metcalfe's law. The more devices they choose to buy, the greater the value of their own network. While many early networking products were initially designed for targeted markets, such as the military or academia, 3Com offered something different. The core notion of its network-centric devices was that they should have a high number of pervasive connections. They were for everyone.

It wasn't always clear if 3Com would succeed, but the team understood that the value of attaching machines to networks would grow as the networks grew—even if that required sharing innovations with the competition. After all, the competition would help grow the total number of people connected to networks, which would grow the value of 3Com's products. This sometimes led to an odd mix of competitive secretiveness and openness. But it was incumbent on 3Com to out-innovate and out-execute the competition, given the even playing field afforded by the standard.

Bob Metcalfe's leadership traits combined with his positive, charismatic style played out well in his early efforts to promote both Ethernet and 3Com. Determination, drive, and a strong cognitive capacity to convey his ideas, and sell them, proved invaluable. His self-confidence and integrity helped establish his open and honest style, one that followers would eagerly trust. While Bob has on occasion been too confident (readers may recall that when he later predicted that the Internet would crash in 2000, and it didn't, he ate his words from a blender), almost forty years later, he has 22,000 followers on Twitter,[14] and is frequently quoted in the press on a variety of topics.

---

13. George Gilder is a remarkable force of nature: an investor, author, economist, and Co-Founder of the Discovery Institute.

14. As this book went to press in 2019, Bob Metcalfe quit using Twitter citing multiple reasons. For Bob's Twitter farewell visit 3comstory.com/tech.

Bob was and is a good example of "trait leadership," exuding traits of high self-esteem, assertiveness, and self-assurance, along with charismatic leadership. Bob had great willingness to share his technology with others who sought to pursue his vision, such as Steve Jobs. And while Jobs may have successfully driven several proprietary platforms for Apple, Metcalfe's passion and ultimate legacy was leading others to embrace an open standard for Ethernet.

Without Ethernet becoming so pervasive, the notion of ubiquitous networking would likely have taken much longer to become realized. And all of the technologies and startups that rely on the availability of free or low-cost networking in their cloud data centers and in WiFi connected networks—such as Facebook, Google, Amazon, Uber, and Salesforce—wouldn't be part of the fabric of our lives today (and wouldn't have allowed the generation of those immense financial successes that eclipsed 3Com's).

"If I were a better person, there would not be a Cisco."

Bob Metcalfe

Howard Charney: VP of Operations—and Social Director

A most critical part of the very early leadership team was Howard Charney. Howard's official role was VP of Operations and General Counsel. But he arguably played an even more vital role as the social glue that fortified the powerful corporate culture which helped 3Com thrive during its early days of explosive growth, and worked in a variety of roles in 3Com's evolution.

Howard was raised in Massapequa Park, on Long Island, New York, not far from where Bob Metcalfe grew up, in the village of Bay Shore. Howard's father worked in the evening as a linotype operator for the newspapers, and his mother raised him with his two sisters. I asked him my question about his formative years: did anything stand out, that aimed him towards engineering, legal, or founding a startup? Howard shared that his father worked evening shifts, and his mother took care of him during the day. In the fourth grade, they called her to the school and asked to hold him back. Howard recalls:

> They said I was a problem child. I wouldn't stay in my seat. I wanted to talk to the other children. I wouldn't take instructions. I was not manageable, and they felt, that in their wisdom, this child needs to be retained. My mother told them, in no uncertain terms, "No way in hell!," and so what happened was over their objections, I suppose, I went on to the fifth grade.

> In the fifth grade I had this teacher, and I'll never forget him. A man who's probably passed on now, but his name was Al Baum, and he taught me the value of paying attention, of trying to learn as much as you could, he taught me the value of being precise and being clear with the use of the English language. I used to marvel listening to this man speak English because his English was something to learn from. What I learned from him actually changed my outlook, it changed my dedication to wanting to study, and it changed the attitude I had about wanting to achieve high grades because before that it was just, "I don't care."
>
> So, it's a funny question that you ask because it isn't exactly, doesn't exactly relate to engineering because I don't think anyone knows, but it tilted my trajectory toward academic excellence and results as opposed to, I think a lot of kids, they don't have any guidance. I don't know if anyone realizes how much that man changed my life, and I don't know if I ever got to tell him that he did.

Howard got his BS and MS from MIT, and an MBA and JD from Santa Clara University, where he has since been a major benefactor to its law school. Howard described what compelled him to leave IBM, where he was a successful attorney, to help launch 3Com, by sharing his speech ideas that he was going to use for Santa Clara University's School of Law commencement ceremony in May of 2018:

> There's this principle in life, it was a concept or defined in this book that I read which said, "If the downside risk is zero, you always take the risk."
>
> Now, in most cases the downside risk is not zero, it's small. But if the negative aspects of what could unfold in front of you are really kind of minor, then maybe you should go off and see if this works. So, when Metcalfe said to me, "I'm going to start this company," and I thought, well, geez, I just passed the bar exam, and I'm a licensed patent attorney, why am I going to do this? And I came to the conclusion that, so what? What's the downside? The downside is in one year, maybe two from now, this enterprise blows apart. It doesn't go anywhere. And then what happens is I go back to being an attorney in private practice. And have I really lost anything?

> Not really. I'm not an old man at that point. I'm just a year or two older and I go do what I was sort of doing before.
>
> So, I took the risk, and I joined him, but 3Com had no concept of what it wanted to do at that time. He knew. I mean, giving him credit, he knew that the specification was going to enter the public domain, from which none of us could make any money because it was going to be donated to the public. The spec. But, he also knew that the spec and the articulation of the spec would create a large market for things which would be part and parcel of that specification. And he somehow in his brain thought that 3Com could be a part of it. But he didn't quite know how but he thought, well, it's gotta work and we'll figure it out.

After leaving 3Com in 1990 (much longer than the year or two he'd contemplated), Howard founded Grand Junction Networks, a leader in Fast Ethernet desktop switching products. He ended up at Cisco in November 1995, after Grand Junction was acquired (ironically a critically adverse event to 3Com's story, as you'll see). He went on to grow Cisco's desktop switching business from $35 million to billions and earned the top market position in switching technology for small and medium-sized businesses. Howard wound up serving as a Senior VP for the Office of the Chairman and CEO at Cisco until January 2018, having held this position since October 1998, an amazing 20-year run following his 3Com and Grand Junction startup years.

But Howard Charney's legal credentials weren't what made him such a valuable part of the early 3Com team. Just about every successful company has that one person that operates like the social director on a luxury cruise, or the bartender at your local pub. At 3Com, Howard was that person. The most common sentiment in interviews with alumni was, "People loved Howard." Chris Paisley told me, "Howard was the glue that held the company together. He was absolutely the person that people rallied around." Chris remembers a time, at his first company meeting with all 250 employees, when Howard was drawing names for a contest that would send ten employees on a vacation to Germany, with Howard himself leading the trip. Chris said, "Traveling with Howard had a ton of appeal to people, and he was the one pulling the names and announcing the winners. You've never seen such joy and exaltation at a meeting, as at that meeting. It was incredible."

Stan Slap, a corporate culture consultant, has analyzed the factors that separate successful companies from those that flounder along. He sums up his recipe for success by observing that a key indicator for success, and a highly quantifiable one at that, is whether or not teams have an emotional commitment towards their managers. In his books and consulting work, he stresses that the emotional commitment often has nothing to do with the company, its products, or its vision to change the world. Often, it's based on something personal, some idiosyncratic connection to one's leaders. If Bob was 3Com's technical visionary, and Bill Krause was the operational guru, Howard was the company's people person par excellence.

Howard Charney and Bob Metcalfe were each the social chairman of their fraternity in consecutive years. Harvard's Joe Lassiter remembers that the young Howard was "very good at working with people, working through problems, and finding solutions." He continued, "Howard's real forte was his ability to interact with people, understand them and organize them to get things to happen."

When Charney moved to San Jose to help IBM innovate disk drives, Lassiter recalls that Howard at the time "was a motorcycle driver and a bit of a rock and roll kind of character." This was such a strong contrast to Bob, who "was your complete preppy kind of guy." Lassiter never quite expected that the two would end up working together, but acknowledged that it was a good marriage of skill sets. "Howard was very thorough and very good at dealing directly with people. Bob was a visionary who could get out in front of an idea and sell it." Bob Metcalfe summarizes Howard Charney with a single word: "Integrity."

Bill Krause: The Making of a President

Bill Krause was born in Philadelphia in 1942 to what he calls two "exceptional parents" who remain his heroes to this day. As a child of military parents, Bill was frequently moved around the globe following his father's career. At age four, he found himself with his mother and infant brother leaving Philadelphia, headed out on a transpacific ship to Japan, where his father was newly stationed. He remembers taking care of his sick mother and brother on the ship at such a young age. (Perhaps this experience imprinted upon him the importance of taking care of everyone during trying circumstances, a talent he'd be called upon to use more than once during his tenure at 3Com.) Although Bill was very young during his days in Japan, it made a strong impression on him. He remembers going to and from kindergarten, living in a tatami house, and interacting with the family's Japanese house staff. In addition to his two years in Japan, other formative experiences abroad would shape his perspective and character.

Bill and his family moved back to Philadelphia when his father was stationed at Fort Dix, New Jersey. He recalls one of his many primary schools,

St. Timothy's. "That's where I got my knuckles rapped with a ruler, from the sister, in third grade." The next year it was on to Camp Pickett in Virginia, where his family moved to a small farm. As a Yankee kid, Bill was often beat up by the farm boys in his grade. One day, as Bill explains, he got tired of being bullied, so "I fought back and got the reputation of being pretty spunky." Bill was able to win the bullies over once they found out he was a good marbles player. Bill attended four different primary schools by fifth grade. This was a challenge, he explains, but it taught him "how to get along with people… how to move into a community and quickly establish yourself, insert yourself into existing groups of friends."

At age 10, the family was off to San Francisco. Bill became an Eagle Scout by age 12, before the family headed off to Bad Kreuznach, Germany. Bill blossomed into an adept athlete. "I was on the football team, captain of the basketball team. I dated the captain of the cheerleading squad." But he admits, "I was a very, very undisciplined youth; growing up as a teenager in Germany was a disaster."

Bill says that it was not until he attended The Citadel as a college student that he started to become the person he is today. He was transitioning from a high school "where there were no rules" to The Citadel "where there were a lot." He recalls that the degree of discipline he experienced at The Citadel "was the best thing that ever happened" to him. The Citadel taught him many lessons, including the importance of perseverance, self-discipline, and follow-through. "Going to The Citadel was hard. When I went there, they had a very abusive freshman system." In later years, as a major donor to the college, Bill has been integral in transforming the school for the better, making it more diverse and inclusive.

Bill Krause's entry into electrical engineering was almost happenstance. On a summer visit to The Citadel, Bill was required to retake the SAT because the school had lost his file. After grading it that same day, Bill was offered a scholarship on the spot. He remembers being asked by the Dean of Admissions, "What do you want to major in?" "I don't know, whatever you think," Bill replied. The Dean continued, "Well, you're good in math and we don't have many students in the electrical engineering department. How about electrical engineering?" Bill received his Electrical Engineering degree from The Citadel in 1963.

Directly out of college, Bill entered the military and at age 21 was responsible for much older non-commissioned officers that were highly technically trained. He ascribed to the notion, "In order to lead, one must learn how to follow." The rigid structure of the military, he felt, taught him how to be a follower, one of the best things he could have learned at the time.

In the military, Bill was taught how to create an "environment of success." He said, "If I could ensure my platoon had a hot cup of coffee in the morning, a good breakfast, a good lunch, clean clothing, a good meal at dinner, and a warm bed at night, I was king."

After military service, Bill wound up at General Electric, and worked his way to the West Coast in 1966, where he had heard from military friends who were having a good time. After a short stint with Technical Measurement Corporation in Los Angeles ended when the company went bankrupt in 1967, Bill then found his way to Hewlett-Packard, where he took on a role that was new to him, and a first for HP: digital sales engineer.

He recalls being thrown in head first. On his first day at the job, Bill was introduced to Bill Hewlett (the "H" in HP) in the cafeteria. Hewlett immediately took an interest in Bill because he would be selling a product that was Hewlett's pet project: the desktop calculator, a forty-pound machine costing $4,900. Hewlett would often call Bill to schedule sales calls together. "I must have made twenty sales calls with [Hewlett]," Bill recalls. Unsurprisingly, Hewlett became one of Bill's long-time mentors, and Bill remembers his time at HP very fondly. He became their youngest ever district sales manager and regional sales manager.

By 1974, Bill Krause was selling the HP3000, when it was in trouble. The hardware wasn't yet ready for the software running on it. Bill helped to push revenues for the product from negative sales (folks were returning the product) to over $1 billion a year. Later in his HP career, Bill was tasked with selling lower-cost computers. He found that selling many of them together—if they were all connected—was more valuable for the customer and generated more revenue for HP. Eventually, Bill discovered Xerox PARC's work on Ethernet. He soon hooked up with Bob Metcalfe, and their first meeting was in Los Altos at Mac's Tea Room, where they discussed the 3Com opportunity.

When Bill was later invited to join 3Com and was mulling over whether or not to leave HP, he recalls asking his wife what she thought. With tears in her eyes, she said, "Why would you ever leave HP? You could be vice president of HP, maybe president. Who knows? Why would you want to leave and join this startup company?" Bill later called Bob Metcalfe, relaying that the conversation with his wife did not go as expected. He thought that a dinner where she could meet Bob might sort things out. After that orchestrated evening, he asked her again "What did you think?" She said, "Well, Howard Charney's one of the smartest people I've ever talked to, and Bob Metcalfe is one of the most charismatic people I ever met. What the hell do they need you for?" So, Bill asked, "Does that mean it's okay?" "Go for it," she replied.

The personal history of Krause's upbringing appears to have had a strong impact on his leadership style. His leadership, execution, decision-making skills (all of which are described in more depth later), as well as the occasional outburst of Bill's short fuse is legend. Perhaps his strength and skills can be traced back to his young development—"making it" in Germany as a teenager, and going on to become a disciplined man in college and military service. In those years, Bill developed skills to survive, as anyone does, based on fate and circumstance. Numerous executives and employees expressed that they had encountered challenges when working with Bill. While Bill had these exceptional strengths, including identifying leadership potential and building strong teams, many people struggled at various times working with him.

Some stories are so absurd that, in retrospect, they become hilarious. One of these became infamous at 3Com, illustrating Bill Krause's short fuse in a humorous light: One Halloween (circa 1988), Bill was upset at sales executive Jerry Dusa during an executive staff meeting, and started yelling at him while wearing his cow costume "head to toe." (By the way, almost all the executives I spoke with insist it was a cow costume, while Bill is quite confident it was a bear costume; sadly, there were no photos of the event!)

Jerry Dusa recalls that the meeting itself was ridiculous, because all senior management had been strongly encouraged to dress up. "We were all sitting around the table in our stupid Halloween costumes having this very heated discussion about millions of dollars' worth of product, orders, and shipments." Bill began to lose his temper during the sales debate, and yelled at Jerry while making vigorous "hand motions through his hooves [or paws, depending who is telling the story]." The Halloween costumes surprisingly did not lighten the mood because, as Jerry remembers it, "You could cut the tension with a knife."

Alan Kessler, who was an integral executive in 3Com's sales and customer support operation—and ultimately helped run Palm as President—related a story about sales executive John Marman and Bill Krause—or, more pointedly, Bill Krause's garment bag. When Alan, John, and Bill had been traveling through Europe doing local reviews, tension among the group was high. John Marman was so angry at Bill that one morning, after Bill had hung his garment bag in a closet and had left the room, "Marman opened the closet and punched out Bill's garment bag." (Seems like this approach to anger management was tailor-made for the occasion.)

This story gained more steam humorously when I spoke to another executive that added, at the time, "our luggage might be crammed in the car, yet Bill's garment bag would be laid gently on top, with special care."

But as Chris Paisley, his CFO, and many others aptly pointed out to me, "If Bill had not been CEO of 3Com in the early years, 3Com would not have achieved the level of success it achieved." Chris continued, "You can talk about Howard and Bob as the founders, but Bill was as dedicated to 3Com as anyone." All three contributed to the magic that made those early years work so effectively.

Chris also observed that while Bill could become agitated over seemingly minor matters, he could also be calm and collected in situations where the potential consequences were far more dire. After seeing Bill ably and supportively handle a major crisis, Chris came to an epiphany which he shared with Bill one day. He told Bill, "I think you are far better when there's a true crisis, and it's something really important, than you are when there's something minor." Chris' observation seemed to ring true, even for Bill who replied, "Well, I believe that it's hard to trust people to get big things right if they can't get the small things right." While Bill could have a tendency to overreact or cause alarm, his intentions were always to seek the best for 3Com.

Debra Engel, whom Bill hired early on to run Human Resources, respected Bill's management expertise. She described his leadership as *operational theater*. "If a decision needed to be made, he'd work the process, and if we couldn't come together, he'd make the decision, it would stick, and it was rarely revisited. We got so much done." She did agree that at times "he could be a challenge."

Dave DePuy (an early sales executive) pointed out some differences between how Bill Krause and Bob Metcalfe worked:

> I liked Krause's directness and his complement to Metcalfe. Metcalfe was about big theory, intuitive, waving of arms or grandiose language, and Krause was much more detailed. "Here are my notes, see I'm carrying my calendar in my pocket." Very specific. We're going to do missions, objectives and strategies, and tactics. We're going to do planning cycles.... Metcalfe and Krause were a great combination because they were so different.

Gerald Petak, 3Com's first division controller, recalled a story about how Bill brought employee recognition to 3Com in memorable fashion. Gerald noted that "early on, Bill wanted to acknowledge individuals with recognition, basically kudos for a job well done. But Bill was unclear about the pronunciation, calling the award a Kadoo. When he announced this at a company meeting, many were flummoxed.... He explained, waited for the head slapping and chuckles to subside, and then made the award. Kadoo stuck and for years, and I think Bill took pride in coining a word with so much meaning at his expense!"

Bill's leadership style, with a strong focus on task accomplishment and goal-setting, is reflected in much of the company's early corporate behavior, which had established structures and production processes. From time to time, Bill's directive, task-focused, goal-setting approach to leadership caused minor explosions from within and without. You could readily place Bill's management style on a leadership grid with two axes: one being concern for people, with the second being attention to production. Bill's mark on the grid would be placed high on the axis for production but, at times, low for people.

Bill reflected on the early years as described in this interview with Michael Copeland of venture capital firm Andreessen Horowitz:[15]

> **MC:** At 3Com you sold networking software that Metcalfe developed initially. Who were the first customers?
>
> **BK:** Bob had developed the first Unix with a TCP/IP stack[16] in it. No sooner had I started at the company when I get a call from this guy in Seattle. It was Bill Gates. He and [Microsoft Co-Founder] Paul Allen were our first customers. They wanted to buy the software to network their [DEC] VAX machines. We spent a lot of time with Gates, and he was always secretive about what he was working on in the back room. Turns out it was DOS [after their purchasing the rights from another Seattle based firm].
>
> Our second customer was a young guy in Cupertino by the name of Steve Jobs. And our third customer was a guy by the name of Andy Bechtolsheim [Co-Founder of Sun Microsystems, and early investor in Google].
>
> **MC:** But it was Ethernet that made 3Com. How did Ethernet beat out competing networking technologies?
>
> **BK:** Ethernet at the time was very kludgy. It was this big, thick cable that at the time you had to screw a tap into, and then you had another

---

15. Copeland, Michael. "40-Pound Calculators, the Birth of Ethernet, and a $100 Billion Mistake: A Conversation with Bill Krause". Andreessen Horowitz. February 2014.

16. Stack refers to layers or foundational protocols that allow for communications on the Internet. The stack ranges from the lowest layer, such as the adapter with its PHY (the physical layer, often called the transceiver) and Media Access Controller (MAC), up to what the user sees at the application layer, such as HTTP, POP, DHCP, etc. For a more detailed explanation of the TCP/IP protocol stack, visit 3comstory.com/tech.

cable that connected to a card that you connected to your computer. The biggest drawback was that and it was about $3,000. Who is going to spend $3,000 to connect a $3,000 computer? How is that to going to work? We knew we needed to get on to a VLSI circuit, a highly dense semi-conductor to quickly bring the cost down, and that was what we ultimately did. So that was one part.

When we did demos we also explained the technology this way: Ethernet worked the same way as human beings having a conversation. We are using the same medium, we are sharing the air. You talk, then I talk. If we start talking at the same time, we both stop, and someone re-initiates one side of the conversation. The reporters got it right away, and understood the advantage over other technologies, but it was Steve Jobs that added a critical piece that led to Ethernet's success.

We had put Ethernet on a card, and instead of having to screw a tap we had a connector that looked a lot like the one you screw into your cable box and TV. And we were all excited about it. We set up four PCs, and we called Steve Jobs who was a good friend and told him, "You have to come over and see this demo." Steve comes over and we hook it up and show it to him.

It was a classic Steve response: "Who's the brain-dead asshole that came up with this shit? This is dreck, this is crap. You want to make it easy to install, just plug it into the telephone jack for cryin' out loud."

Why didn't we think of that? No one knows to this day that Steve Jobs deserves the credit for creating Ethernet the way it is today, and it is a part of why it beat out other competing technologies. It was another one of his brilliant insights around user-interface.[17]

We were also very lucky that about the time we came up with this little card that you could put in a machine, HP came out with the

---

17. It's a bit of an oversimplification to say Jobs was responsible for the advent of twisted-pair Ethernet. Twisted-pair Ethernet and Token Ring were first developed and brought to market by SynOptics Communications in the late 1980s, and not standardized as 10Base-T until much later in 1990. According to Rich Siefert, who demonstrated the first thick-wire Ethernet to Ken Olsen of DEC in 1980, Ken said he also wanted something like telephone wire for deploying computers in schools. Ron Crane also had earlier invented a twisted-pair solution but lost out in the standards war.

Laser Printer and Apple had the Mac and its Laser Writer. These laser printers were $6,500 or $7,000 a printer. Wouldn't it be cool if you could share your expensive printer? To make the math easy, if you ginned together 10 PCs, you got a printer for $600. That is what jumpstarted things for Ethernet. And once email came along things really started to snowball.

MC: You made Ethernet both easy to use and easy for people to buy. Was selling networking gear on store shelves an obvious decision in the mid '80s?

BK: One of our key strategies was how we defined our customers, which were those people who bought PCs. What that meant was that our products would be sold where PCs would be bought, and at the time it meant computer stores. That was heresy. All our venture capital investors thought if you were going to sell networks it had to be done through the IT department. But we fought that. We were determined to sell our products at computer stores because they were easy to install and use. That turned out to be a big success.

MC: All right, enough victories. As CEO at 3Com what bad calls did you make?

BK: You could say I made a $100 billion mistake. The whole world had evolved around networks of PCs. But in the early '90s along came the concept of networking networks—i.e., a little company called Cisco.

We had acquired a company called Bridge that had competing technology to Cisco, but we never could really get it front-and-center because we came from this network of PCs orientation, not networks of networks. As a result of our slowness to get to the market, Cisco got a chance and got born. And I made a $100 billion mistake.

What that showed me is that innovations aren't simple, plain and obvious. Sometimes you have to be hit over a head with a 2×4 to recognize that the world is moving on.

Chapter 4: Wiring the Future

In the late 1970s, as Metcalfe, Charney, and Krause were emerging as professionals-to-watch in the tech sector, the world was undergoing some key changes that would frame the next steps in their careers. While the United States was being challenged in 1979 by the Iran hostage crisis, and was facing double-digit inflation, as well as formidable competition from the Japanese in the auto, steel and TV industries, Silicon Valley was beginning to be noticed. High technology employment expanded 77 percent between 1974 and 1980. New industry groups, such as the National Venture Capital Association founded in 1977, sent some of Silicon Valley's top execs to Washington to explain the importance of microchips, biotech, PCs, and software. And due to a regulatory change in 1978 that permitted pension funds to consider venture capital investments, "prudent" money began pouring into startups. In 1980, there were 47 venture capital firms and was up to 113 by 1983. By 1983, the total investment by VC firms had nearly quintupled, from $2.5 billion to $12 billion.[18]

With far more money to spread around, the 1980s were heady, giddy, go-go years for Silicon Valley. Apple's initial public offering took place in December of 1980, and other notable IPOs followed in quick succession. Computer Associates (now called CA Technologies) went public in 1982, the same year *Time* Magazine named the PC "Man of the Year." 1983 saw the arrival of LSI, Compaq, Lotus, Businessland, and Ethernet competitor Ungermann-Bass[19] as public companies. Also in 1983, Microsoft released Word, followed by Windows in 1985. In 1986, Oracle, Sun Microsystems, Cypress Semiconductor, and Microsoft all became public companies, followed by BMC and SynOptics in 1988, Cabletron and Electronic Arts in 1989. Cisco went public in 1990, and Wellfleet in 1991. A new gold rush was in full swing. Companies and technically minded employees were eager to stake their claims.

While venture investment was booming, there was a simultaneous shift away from funding research into pure science and technology. During the Reagan administration, ARPA became more narrowly focused on specific defense projects. It was renamed DARPA (with the D standing for Defense) to make that focus clear. Many universities were increasingly motivated by financial returns on their research investments, rather than by traditional academic notions about

---

18. In 2017, there was $84 billion invested in 8,076 deals, and 209 VC funds were closed in 2017, aggregating $32 billion of capital to be committed over the fund's life. "VCs invested the most capital in 2017 since the dotcom era". Venture Beat. January 2018.

19. U-B was later acquired by Tandem in 1988, and subsequently sold to Newbridge Networks in 1997, which laid off much of the company. Charlie Bass was also a seed round investor in Synernetics in 1988, the successful switch company 3Com bought in 1994.

the value of learning, and the desire to share knowledge for mutual benefit. In the 1980s, as computer networking was taking off, the division between academia and entrepreneurialism became blurred. As a result, academic research became more narrowly focused on arenas that could quickly be moved to the commercial marketplace, with less emphasis on longer-term research.[20]

Against this backdrop, tech entrepreneurs were planning their futures—and ours.

3Com's initial success was a one-two punch—first, helping to invent Ethernet technology and second, obtaining Xerox's permission to use the patent while simultaneously convincing the DIX alliance to promote the open standard[21] Ethernet as the basis for commercial products. At the same time, Ron Crane, 3Com employee number four, played a huge role in designing the first 10 megabits per second (Mbps) coax-based Ethernet and standardizing this technology via the IEEE 802.3 standards group, which achieved approval in 1985.[22] Rich Siefert, at the time working at DEC, also helped with this work, and pointed out that Ron later helped develop the thin-coax version of Ethernet, solving the reliability and installation problems of connecting devices to thick-coax cable. And as several people related to me, Bob Metcalfe relied on Ron Crane in those early days in much the same way Steve Jobs relied on Steve Wozniak. And like Woz, Ron had a quirky personality that added enormous value to the company.

During these early years, there was an unusual sense of technological community. 3Com was a fierce competitor, but it also helped many other tech companies along the way. As Ron Crane is quoted as saying in Urs von Burg's *The Triumph of Ethernet*:

> We saw our growth linked to the growth of Ethernet as a whole. Anything that made Ethernet look bad would hurt us. So, it was in our interest that all of our competitors at least got their interfaces right so that the standards worked out. If the rest of their products failed, that was fine, but we did not want their product to fail because

20. Estrin, Judy. "Closing the Innovation Gap: Reigniting the Spark of Creativity in a Global Economy". McGraw-Hill. 2009.

21. There have since been many other successful open standards: HTTP, HTML, WAP, XML, and SQL. These are typically built by software engineers, from IT or software firms who collaborate under the auspices of non-commercial organizations, such as W3C or IETF. Many successful open source software projects are available for use by anyone, including Linux, Apache, Mozilla, and my personal favorite, Hazelcast, an open source in-memory data grid solution used where I worked until retiring in 2014.

22. If you want to know more about the local area network (LAN) and wide area network (WAN) products and the Transmission Control Protocol and Internet Protocol (TCP/IP) layers of technology, you can find an overview of some of the basic networking technologies (hubs, routers, and switches, etc.) at the technicalities section of our book site at 3comstory.com/tech.

> Ethernet failed. As a result of this collaborative spirit, the Ethernet community was not merely a techno-economic phenomenon based on competing firms' adherence to a technological standard, but it was also a social phenomenon because of the dense web of personal and business interactions that grew up around that standard.[23]

With this new and evolving technology, 3Com set out to make products that could fit into the existing marketplace. First, for the minicomputer market, they shipped UNET software for UNIX computers in 1981, and later that year, they began shipping the Q-Bus Ethernet Controllers for DEC minicomputers called PDP-11s, as well as Unibus Ethernet adapters used in VAXs.

3Com's DNA

Even as the company expanded, 3Com retained several distinctive characteristics throughout its early years. One of these characteristics was not taking *no* for an answer. Dave DePuy, a U.C. Berkeley MBA grad in 1981, initially applied to 3Com for a job in marketing but was told, "We have one of those," referring to David Coulson. DePuy then asked Bill Krause, "What jobs do you have?" They needed help in manufacturing, so Dave volunteered for the job, then moved to engineering, and later moved to the East Coast to head up their early federal sales business.

DePuy also highlighted another distinctive 3Com trait, a tendency toward audacity:

> Metcalfe had used his contacts at DEC to allow us to do testing on DEC equipment, back in their test labs at night. So, when their testers would go home at say five or six, we'd show up at seven or eight and test all night. This would allow us to say that our products were compatible and tested with the VAX family of computers including the hot 750. We put together the Q-Bus Ethernet adapter (three boards with a jumper cable joining them), and the Unibus controller in November of 1981, followed by shipping a Multibus adapter, by April of 1982 to Sun, Wicat, and others.

Something else that would become a core part of 3Com's makeup was its sales strategy. In February of 1983, 3Com faced a dilemma. Should the company continue to sell directly to customers, or should 3Com embrace the recently

---

23. von Burg, Urs. "The Triumph of Ethernet: Technological Communities and the Battle for the LAN Standard". Stanford University Press. 2002.

developed value-added reseller (VAR) model and the hot retail channels being created by chains like ComputerLand and Businessland? These channels offered a very cost-effective way to reach millions of customers quickly at the cost of larger discounts. The dilemma was resolved at a long weekend offsite meeting (led by Mike Halaburka, 3Com's first VP of Sales) at which it was decided that 3Com would attempt to sell through resellers, if possible. How 3Com selected its method of distribution, and early product lines, would have a profound effect on the years to come.

Another aspect of 3Com's DNA was to never rest on your laurels, or accept a solution that was "just good enough." Whether a product was headed for retail distribution or for direct sales to large OEM customers, there was often a tension between pushing the innovation envelope just a bit more versus making the cash register go "ka-ching" just a bit sooner. In 1982, Bob Metcalfe was trying to convince Ron Crane, who could be a bit of a distracted engineer, to focus on the release of their first Etherlink card for the PC. He had delayed it over concerns about possible failure resulting from a lightning strike. At Ron's memorial service held at the Computer History Museum (sadly, he passed in June 2017), Bob related this story in detail:

> It turned out he was working on a lightning protector to the circuit to put on the part. So, I said, "Ron, here's the product spec. The word lightning does not appear on this spec. Nobody wants lightning protection for this Etherlink. We've never had a customer complain. Ron, drop it! Release this thing to manufacturing."
>
> Well, Ron would not drop it, so he continued working on the lightning circuit…
>
> We sold a thousand of these new Etherlinks to a big bank in New York, and this bank was shrewd, so they bought a thousand from us and they bought a thousand from one of our competitors. And then they installed it, and they had the network running, and what do you all think happened? Lightning struck. And our competitor's cards were all fried, and our cards sat there just humming away happy as clams. And Ron could not stop smiling, vindicated of his stubborn attachment to this lightning circuit. And the customer placed an order for another 1,000 cards.

The Etherlink product eventually shipped millions per month and was a critical factor in allowing the company to go public. A complete version of

Bob Metcalfe's remarks at Ron Crane's 2017 memorial service is included in Appendix A.

Chapter 5: Making Miracles

The 1980s started with a "Miracle on Ice" by the U.S. Olympic Hockey team defeating the Soviets in Lake Placid, Jimmy Carter moving out of the White House amidst the Iran hostage crisis, and Ronald Reagan moving in, on a platform of lower taxes, less government, and a strong national defense—a direction that would later help 3Com's federal sales efforts. In addition to the conservative swing in political mood, it was a time when technical miracles were seemingly a dime a dozen.

In 1981, IBM announced the personal computer, with 16KB of memory, floppy disk drives, and an optional color monitor, all running on the MS-DOS operating system. Consumer electronics such as CDs, VCRs, camcorders, and fax machines became almost required to own. And a year later, Compaq was incorporated and shortly became the first company to legally reverse engineer IBM's PC BIOS (the firmware that would allow them to create compatible PCs). They later became the largest supplier of PC systems in the 1990s. In 1983, something we now take for granted, the Global Positioning System (GPS) was opened for use by civilian aircraft, thereby launching the trend of geographic mapping tools. Around this time as well, early versions of cellphones were introduced. And over the course of the 1980s, the graphical user interface would rapidly move from being a clever curiosity to becoming the only game in town.

With technical miracles like these becoming commonplace, commercial funding for the next wave of miracles became more abundant. In the fall of 1980, Bob Metcalfe met with venture capital firms to raise money. His pitch was built around a vision he sketched out one day.

This vision was a good description of the tech landscape in 1980, and a remarkably prescient and accurate view of the landscape that would unfold over the next few decades. Bob and his small team of eight[24] had effectively

24. The 3Com team in 1980 consisted of another director (Paul Baran, the inventor of packet switching), and employees Bruce Borden, UNET Product Manager, Howard Charney, VP of Operations and General Counsel, Ronald Crane, Director of Engineering, Robert Metcalfe, President, Kenneth Morse, VP Marketing, Gregory Shaw, VP Software Engineering, Sarah Shevick, Secretary, and David Spiller, Director of Finance/Treasurer.

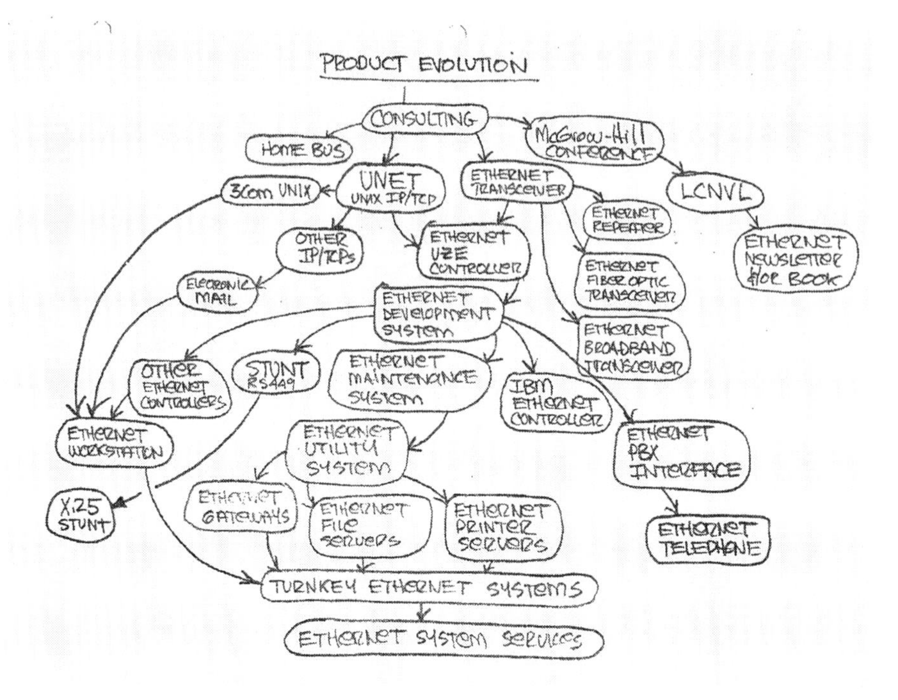

identified the "probable futures" for the marketplace. This was documented aptly by Bob on this slide in his VC pitch deck (handwritten after the rest of his deck was typed). His vision foresaw a vast landscape that has now indeed come to pass, with "miracles" that included Ethernet workstations, electronic mail, Ethernet gateways, controllers, file servers, printer servers, transceivers (Ethernet, fiber optic, and broadband), and Ethernet telephones.

Metcalfe backed up this plan of the future with this marketing strategy:

MARKETING STRATEGY

0 FOCUS ON DEC INTEL XEROX ETHERNET COMPATIBILITY (+HP+IEEE+EIS+IBM...)
0 OFFER BOTH HARDWARE AND SOFTWARE TO MAXIMIZE VALUE ADDED
0 SELL TO COMPATIBILITY ENGINEERING MARKET TO POSITION FOR LATER
0 EXPLOIT PATENTS, COPYRIGHTS, TRADE/SERVICE MARKS, AND TRADE SECRETS
0 RETAIN PREMIUM QUALITY IMAGE, EMPHASIZING RELIABILITY, AVAILABILITY, SERVICEABILITY
0 POSITION 3COM AS A CERTIFICATION MARK TO SUSTAIN PREMIUM PRICING
0 EMPHASIZE CUSTOMER EDUCATION FOR MARKET DEVELOPMENT
0 SELL OEM TO (1) ETHERNET ENGINEERS AND (2) ETHERNET INTEGRATORS, THEN (3) USERS.
0 DIRECT SALES FORCE BACKED UP BY STRONG CUSTOMER ENGINEERING
0 SELL TRANSCEIVERS TO ESTABLISH CREDIBILITY, BLACK ART EARLY
0 PUT SOFTWARE IN CONTROLLERS TO ADD VALUE, IN PARTICULAR PROTOCOL COMPATIBILITY
0 RUSH TO DEVELOPMENT SYSTEM, THEN TIME SUBSEQUENT PRODUCTS WITH SECOND WAVE
0 DEVELOP INTERNATIONAL SALES EARLY
0 USE IP/TCP STANDARD FOR EARLY DOD SALES, LATER COMMERCIAL COMPATIBILITY
0 PHASES: COMPATIBILITY ENGINEERING, SINGLE VENDOR SYSTEMS, MULTIVENDOR SYSTEMS
0 SERVE SECOND SOURCE ROLE FOR XEROX AND DEC
0 USE UNIX AS SOFTWARE COMPATIBILITY STANDARD

17.

Of course, the benefit of hindsight makes everything obvious. But the venture capitalists involved recalled how uncertain things seemed for 3Com at the time. "The company had come together," stated Wally Davis, one of the founders of the Mayfield VC firm, "largely because of Bob Metcalfe's renown and his technical ability." Davis recalled that nobody believed his optimistic business plan. Another investor, Dick Kramlich, general partner at New Enterprise Associates, recalled that no one knew exactly what kind of market 3Com was going to be in back then, Dick noted, "So we didn't know what kind of business guy we would need to oversee the operations of the company. It was clear, however, as Bob's own instincts suggested, that by the summer of 1980 we needed a little adult supervision."[25]

Before leaving Xerox, Bob had done extensive research on entrepreneurial ventures. He had identified three recurring mistakes that often afflicted startups at the time (and still do today): letting the founder's ego limit the growth of the company, lacking a focus for the company, and being insufficiently capitalized for growth. 3Com was facing the third problem in early 1981. To remedy this, they sought not only financing but also expertise. "We sold our shares at a discount to attract premium venture capitalists," Bob said. "We wanted people who would give us more than money."

Part of the venture capital arrangement that they settled on was an agreement that Bob would recruit a president for the company. In the aftermath of this $1.1 million funding round for one-third of the company, Mayfield VC Wally Davis introduced HP executive Bill Krause to Metcalfe. Bob first talked with Bill Krause at Mac's Tea Room in Los Altos in January of 1981 and Bill joined 3Com as President of the company that March. He was employee number 12.

Bob Metcalfe retained his CEO and Chairperson role. Bill received nine percent of the company, second only to Metcalfe's 21 percent. Shortly after, Krause hired a VP of Sales (Mike Halaburka), a VP of marketing (Larry Hartke), and, a few months later, a VP of Engineering (Larry Birenbaum), who took over engineering for Ron Crane and Bob Metcalfe. In 1982 the board moved Bill to CEO and Bill moved Bob Metcalfe to Chairman, VP of Marketing, until they went public in 1984. Bob Metcalfe remained Chairman for the next eight years, until 1988.

Gerald Petak recalled, "Krause was the business leader and Metcalfe was the Technology visionary. So much so that Metcalfe was consulting with GE (General Electric) to gain some starter money, in addition to the revenue for the Q-Bus product from DEC. The GE consulting work was on a product named the Home Bus, a networking scheme to connect home appliances." Imagine, in 1981, thinking about the Internet of Things or a device like a

25. Richman, Tom. "Who's in Charge Here?". Inc. Magazine. June 1989.

WiFi thermostat like Nest, before the PCs were even networked! Gerald also recalled a moment with Bill that impressed him, and demonstrated his charismatic and self-confident style. With a crisis on his hands, Bill conveyed to the team his plan to fix it, and told them, "If any of you are not confident, about this, I am. I have enough confidence for all of you!"

Bill recalls his first conversations with Metcalfe and how Metcalfe shared the idea of there being "three kinds of work: hard work, smart work, and team work." This principle was verbalized and written about frequently within the company throughout its early days. Bill and the executives also encouraged the idea of "confederated entrepreneurship," with its cornerstone of ownership of stock by all employees. Confederated entrepreneurship was often described in Bill's notes to the company as "teams of people operating with a great deal of independence and freedom, bound together by a common company mission and objectives."

With "adult supervision" in place and some cash in the bank, 3Com's sales would soon be gliding up the proverbial hockey-stick curve. As Bob noted, "Having experienced people doing important jobs makes it a hell of a lot easier." A second round of $2.1 million was raised in January of 1982, and a third round of $3.6 million came in July 1983, with the help of some investment bankers who anticipated the company going public. This brought the total investment in 3Com with earlier rounds to $6.8 million. Bob remembered, "Orders grew 23 percent per month for seven months in a row, and 25 percent per month for five months in a row. The EtherSeries products were exploding. The first six months ending November 1983, were $6.1 million, compared with the six months that ended November 1982, which were $1.8 million."[26]

However, that success would only come after Bill Krause established himself at 3Com by navigating through some turbulent growing pains.

Ship Happens

In January 1981, 3Com shipped its first commercial product: UNET software for networking UNIX-based computers. Subsequently, the first Ethernet transceivers were built and packaged for shipment in March 1981 led by Ron Crane's design, with the help of Tat Lam, a contract worker who later established his own transceiver company called TCL, with his resourceful grandmother, who helped wind wire around transformers in his garage in Fremont, and a team of employees working in a stifling hot locker room turned "production" facility. Tat's grandmother was a gifted seamstress, and winding these toroidal transceivers took skill and patience. These transceivers sold for $650, or $1,000 for a higher performance model. Today, high-performance adapter chips that

26. Electronic Engineering Times. "Robert Metcalfe of 3Com". April 1984.

perform the same function are less than 50¢. (I use the word adapter loosely, as 3Comers often did; the more proper name is network interface card, or NIC.)

Epoxy potting compound was poured into the transceivers to provide mechanical rigidity, hide the technology from competitors and, as some have suggested, make the customer feel that they were buying something of great value and weight. Anyone at 3Com in the early days undoubtedly heard the story of how the staff had to manually mix the 2-part epoxy. The staff received mixing instructions, and encouraging reminders from Howard Charney, who would chime, "Stir and stir the epoxy until it looks done, then stir some more!"

A little epoxy, however, wasn't enough to keep the balance sheet in the black. In the summer and fall of 1981, 3Com was losing money. A "Four Month Survival Plan" was put into place to keep the company afloat, and a hiring freeze and expense controls were instituted. Even in the early years, management was disciplined about fiscal control to keep the business healthy.

The objectives of the plan also included meeting profit and cash flow targets, an order rate of $200K a month, and expanding public relations efforts. At this point, Bill was also taking steps to bring Mike Halaburka, the new VP of Sales, up to speed, close distributor agreements, book orders with DEC, finalize OEM agreements with Xerox, Intel, and DEC, and manufacture these products all while promoting their accomplishments to the press. Bill was also intent on creating broader awareness of the company's pricing and program licenses.

Soon after coming on board, Bill Krause developed critical business planning processes with the management team, which included an annual process for developing strategy and tactics, and the creation of five-year plans. Each year, a Fiscal Year Operating Plan was published and distributed around the company; it included the famous MOST approach: Mission, Objectives, Strategies, and Tactics. Bill was arguably one of the most goal-oriented executives in the valley. He demonstrated charismatic, behavioral, and contingency styles of leadership, with decision-making bolstered by a focus on structure, tasks, and setting goals, Bill would say, "All things in three," and lay out three critical things to accomplish in the Operating Plan, such as: get gross margins up, get revenue up, and control spending and hiring.

In September 1981, company operations moved from the smaller 3000 Sand Hill Road facility in Menlo Park to a 14,000-square-foot facility at 1390 Shorebird Way in Mountain View. This provided the opportunity for 3Com to create its first "real" manufacturing facility, an upgrade from the prototype operations at Sand Hill Road. Apparently, the floors at the Shorebird site were originally all carpeted; the company had to exchange the carpet for tile, to impress customers with a more legitimate "manufacturing" look (as well as to help remove risk of electrostatic discharge from the carpeted manufacturing and lab spaces).

In October 1981, 3Com made its first shipments of Q-Bus Ethernet (QE) controllers for networking LSI-lls, DEC's PDP-11 minicomputer. Uniquely, this minicomputer contained the entire CPU on four LSI chips made by Western Digital.

In the 1980s, the IBM PC and its clones took over the small computer market. Newer processors like the Intel 8088 used in these PCs could outperform the old DEC machines. This created an opportunity for 3Com to build a non-custom hardware controller that could be combined with the transceiver and cables and sold to customers as a "starter kit" for connecting office computers.

In 2004, when 3Com was celebrating its 25th anniversary, Bill shared a few thoughts about these early years in an interview with Karenda Botelho, who worked in 3Com's employee communications from 1994 to 2004:

> **KB:** What was it like being CEO of 3Com in its early days?
>
> **BK:** In short, it was a frenetic combination of creating a company "built to last" ... and trying to survive by frantically prospecting to find customers who might have an interest in connecting their computers via Ethernet.
>
> First, some context to set the stage. While Bob Metcalfe and I got the most visibility as 3Com's early leaders, Howard Charney was a key and essential member of the triad that led 3Com during my days as President & CEO. When I joined Bob and Howard to start the journey we called 3Com, there were two key goals that we had in common: (1) Given that Bob came from Xerox, Howard from IBM, and myself from HP, we aspired to build 3Com into an institution that would withstand the test of time, live beyond its early founders and become an icon in the networking industry just like our respective companies had done before us and (2) if we were lucky to achieve the goal just outlined, we would all retire after about 10 years of service so that 3Com would gain the benefit of a fresh set of ideas from new executive leadership—we felt that executives had life cycles just like products do.
>
> So, given our key goal #1, we set off to create a company "Built to Last" even though the famous book on this topic was not written until some 13 years later. We consciously worked on statements of vision, values, mission, strategy, and day to day tactics and widely

> communicated these statements frequently—some might say too frequently. As part of our vision, we created a measurable, easy to understand, long term goal for the company, "Ethernet One Million PCs by the end of the decade (1989)." This type of statement, later became popular in the business press by the term "BHAG," which stood for "Big Hairy Audacious Goal".
>
> We knew that one of our competitive advantages was to be able to move faster than our larger competitors. We believed that if everyone was on the same page with regards to these key statements, then we would get the leverage of each individual acting quickly and making decisions on their own, but in line with our common goals. We even created a term for this called "Confederated Entrepreneurship."
>
> Now with this context in mind let me answer the question of what was it like being the CEO of a young, growing company. It was a mixture of euphoria and panic. Given the early culture that we had at 3Com, all 3Comers were tuned in, turned on and passionate about using our Ethernet technology to create a paradigm shift in the way computers communicated compatibly. However, we were panic-stricken about whether or not we could find enough customers for our early products to survive financially.
>
> So, in these early days my job was twofold: (1) be the company's chief cheerleader during the days when panic prevailed and (2) be the company's chief devil's advocate when we got too euphoric over our successes.[27]

In the same interview, Bill noted that the three most essential decisions were, first, leveraging Ethernet, not Token Ring or other LAN technologies, and second, selling through distribution channels, such as PC retail stores and newly formed LAN system integrators, and third, investing to be the lowest cost supplier. These three strategic decisions were huge risks that paid off and became the cornerstones around which 3Com grew into a multibillion-dollar enterprise.

Leveraging the channels where PCs were bought was hugely successful—ComputerLand, Businessland, Entre, Morris Decisions, etc. It helped squash competitors like Nestar, Corvus, and Sytek. The conventional wisdom at the time was that selling complex products, such as local area network products,

27. Botelho, Korenda. Interview for 3Com's 25th Anniversary Party. 2004.

could only be done through a direct salesforce; selling LANs through PC retailers was unheard of. And, at the time, Ethernet was far from a ubiquitous standard; it was a risky bet when compared to IBM's competing and dominant Token Ring. How risky? To paraphrase the old saw: "Nobody ever went broke betting on an IBM standard."

The first shipment of Unibus Ethernet (UE) controllers was dispatched in November 1981. This hardware product provided for connections between VAX systems. The company then raised a second round of financing in January 1982 amounting to $2.1 million, and Multibus adapters, shipped in April 1982, went to out the door for customers Sun Microsystems, WICAT, and Convergent Technologies.

In the summer of 1982, a quality problem was discovered with the UEs and shipments were put on hold for six weeks until the problem could be corrected. This was a difficult decision to make. The company had grown to around thirty people, so anything that jeopardized revenue also jeopardized paychecks. But, the priority was placed on quality and "not shipping crap."

Following that setback, 3Com would soon rack up a big hit. In 1982, it would become the first company to deliver Ethernet connectivity for the IBM Personal Computer. IBM's first PC—the personal computer that was actually called the Personal Computer—surprised many who were skeptical that IBM could develop a competitive product in this fast-moving entrepreneurial market. But IBM's $1,565 16-bit machine soon dominated what had previously been called the microcomputer market.

3Com's sales took off after introducing the first Ethernet product for the IBM PC. It was called the IE (later branded as Etherlink) and it used integrated circuit technology from Seeq Technology to combine a controller and transceiver on one board. The first several months' shipments were unit volumes of a few hundred. At this time, 3Com also introduced a network server as part of the EtherSeries family on an Altos computer platform, which provided print-, file-, and mail-server functionality. After the board named Bill Krause as the CEO in June 1982 and put Bob Metcalfe in charge of Sales and Marketing, Metcalfe went on to help get the company up to $1 million a month in sales, based on his credentials and contacts, which was a critical success in those early days. As PC shipments exploded, so did 3Com sales.

1983 saw the rapid growth of the Ethernet adapter and transceiver business. And Seeq's Ethernet chip, with the first to use very-large-scale integration (VLSI) technology, was a huge and important factor in driving down Etherlink's

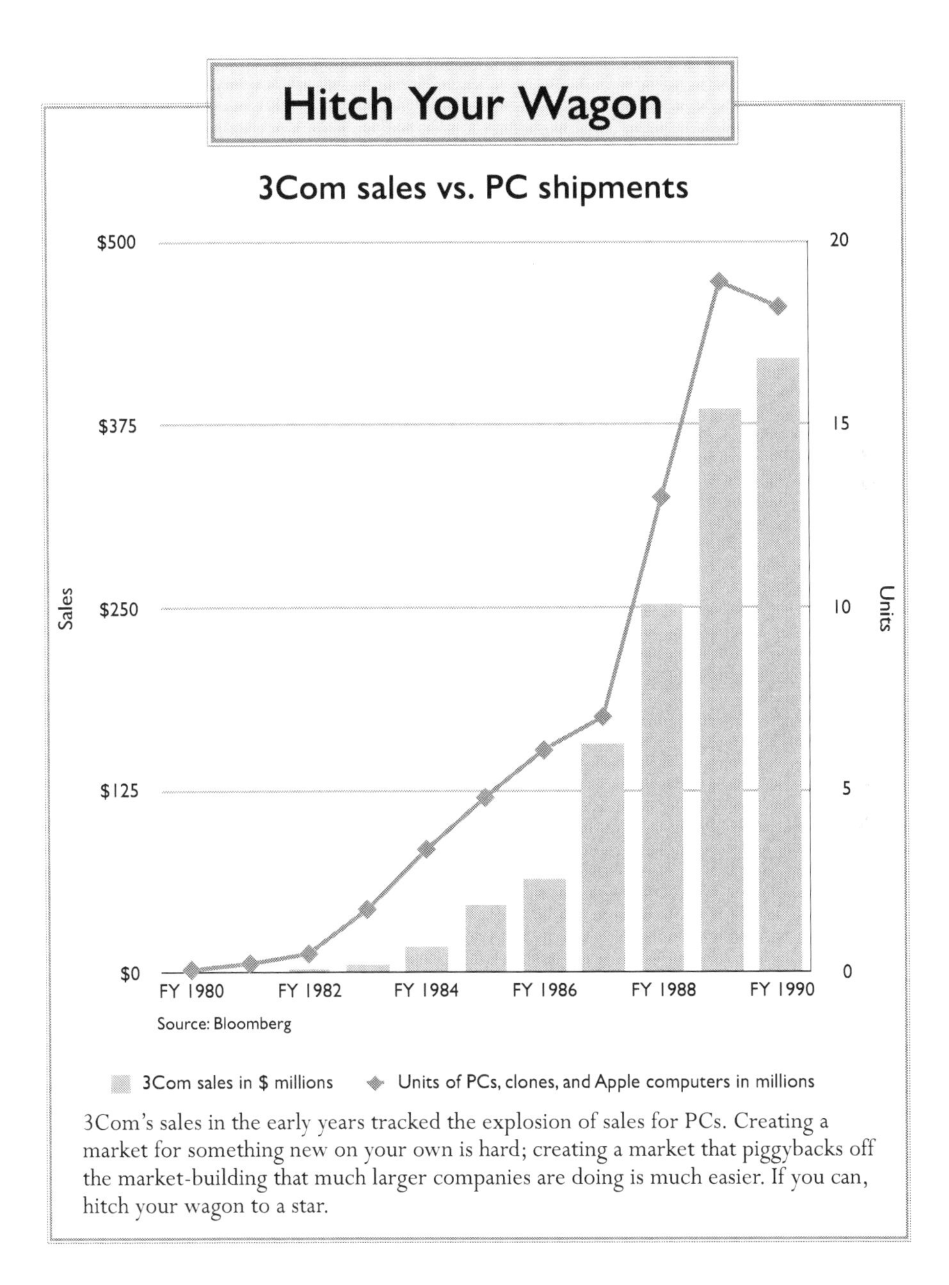

3Com's sales in the early years tracked the explosion of sales for PCs. Creating a market for something new on your own is hard; creating a market that piggybacks off the market-building that much larger companies are doing is much easier. If you can, hitch your wagon to a star.

cost below $1,000, and enabling 3Com to fit all the electronics onto one board. As a joint effort of the two companies, the VLSI chip paid off, and made 3Com a first-mover in the PC connectivity arena. It also included software that enabled it to print VisiCalc documents across the network without exiting the application program—imagine![28]

"One of the great moments in my career there came early as a product manager in the mid-1980s. Steve Jobs desperately wanted to connect his Mac to the "Internet" and Derek Brown and I were assigned to make that happen. I can also recall being invited to lunch with Bob Metcalfe, Bill Krause and Howard Charney to discuss whether we should invest any time and energy on building an adapter for the Mac! [they did!]"

Elaine Hanson, Former President, 3Com Canada

Expansion—and learning to manage growth—was the order of the day. The company settled into the full 22,500 square feet of space in their new Mountain View location. To keep up with growth, the team worked many weekends, and often found themselves grilling hamburgers or ordering in pizza for lunch. From 1983 to 1984, sales grew from $4.7 million to $16.7 million, which translated to over 250 percent growth. According to Gerald Petak, a computer on the desk of every employee became a reality in 1983. "Krause's tactic was to have 3Com be its own best customer, to use and improve the very products they were selling. As some said: one big ass demo!"

Total staffing demonstrated comparable growth, up from 48 to 136. Debra Engel came on board as VP of Human Resources (later known as Corporate Services), along with Charles Kempton, as VP of Sales and Marketing. The EtherSeries products were enhanced, and specialized versions were launched for personal computers from Texas Instruments, IBM, and HP. At the time, 350 dealer locations were trained to sell EtherSeries products. The Etherlink itself had also become more affordable; its price was reduced from $950 to $795. At this time, the board consisted of Bob Metcalfe, Bill Krause,

28. PC Magazine. "Ethernet comes to the PC". January 1983.

Paul Baran[29] (Cabledata Associates), Stephen Johnson (Komag), C. Richard Kramlich (NEA), Jack Melchor (Melchor Venture Management), and F. Gibson Myers (Mayfield).

For a presentation in 1983, Bill published a note on the fundamentals of 3Com corporate culture that included the following:[30]

- Feedback
    - Department communication and evaluation of monthly and annual plans
    - Management by priority
    - Yearly written performance reviews
    - Access to the network—no hidden agendas
    - Monthly department priorities
    - Financials
    - Executive Committee minutes published
- Hiring and compensation
    - High quality—high salary
    - Productivity and personnel turnover measured
    - Salary supplemented with stock options
- Open information flow
    - Access to the network
    - Monthly one-on-ones
    - Weekly coffee and donuts
    - Management by walking around
    - Open workspace

Debra Engel, recruited by Bill in 1983, stayed on until after the US Robotics merger in 1998. Debra had HP in her blood and helped to evolve 3Com's culture largely in its image. HP and 3Com were workplaces known for being open, respectful, consensus oriented, confrontational when necessary but

29. The impact of Paul Baran's professional contributions cannot be overstated. In the 1960s, he led the development of techniques to send data in small packets through a network, while working on Cold War issues for the RAND Corporation. In 1969, those concepts provided the impetus for the Department of Defense to create the ARPANET, the precursor to the Internet. Paul started CableData Associates, Equatorial Communications, Telebit, and Packet Technologies/StrataCom. In 1986 he co-founded Metricom and Ricochet wireless (I interviewed with Metricom in Los Gatos in the 1990s). A wonderful interview on Paul's work at RAND is available courtesy of the Computer History Museum.

30. My experience is that Silicon Valley startup CEOs sometimes give short shrift to these kinds of culture setting routines, and often ignore the structure of annual employee reviews, written priorities, and other process and structural discipline, but at their own peril.

driven by good ideas over politics. Also, in 1983, Bill hired John Celii, their first CFO, a charismatic, seasoned HP finance executive. John later worked for Convergent Technologies, Grand Junction Networks, and Cisco, and has been involved in the horse racing scene. One of John's first moves after taking 3Com through its IPO, was to hire Cindy Hawkins from Deloitte, the firm that took them public in 1984, who became 3Com's first corporate controller.

Keep Calm and IPO On

In 1984, 3Com became a publicly traded company, trading under the symbol COMS. The company's initial public offering of stock opened at $6 per share. By June of 1989, the stock traded at $28 with a market capitalization of $786 million, while Cisco was still a privately held company. This was an exciting time for everyone in the company, particularly since all employees had stock options. Bill Krause told the story of what it was like when 3Com went public in the same interview with Karenda Botelho:

3Com Executive Committee (later called ExecCom), Source: 1984 Annual Report. Seated left to right, Bob Metcalfe, Bill Krause, Howard Charney, standing left to right, John Celii Jr., Debra Engel, Charles Kempton, Dan Robertson. Larry Birenbaum not pictured.

> **KB:** What was it like to be CEO when 3Com went public?
>
> **BK:** The same feelings of euphoria and panic [as running a company]. We were euphoric about the exciting experience of going public and at the same time panicked about the responsibilities of being a public company. However, even though going public was a very satisfying experience, we viewed it calmly as another key milestone and talked about it as the beginning of phase two of 3Com's development as a company.
>
> When we went public in 1984 it was one of the most difficult times to go public in the early history of new technology companies. It was as difficult to go public in 1984 as it has been to go public in 2002-2004.

As an example, our investment bankers started out telling us that we would go public with a stock price of between $10-$12. On the eve of going public, we were told that the price would be $6 and, even at that price, the offering would be very difficult to complete. Talk about panic, especially after having invested considerable time and money in the process. We felt we had no other choice but to proceed. Fortunately, with the benefit of hindsight, we made the right decision.

Moving—and Moving Onward and Upward

Two things inevitably follow a successful public offering: first, the company finds itself with plenty of capital to invest in its future, and second, its previously private goings-on become subject to public scrutiny. And actually, a third thing usually happens: employees who may have never paid attention to the stock market before suddenly find themselves checking share prices multiple times a day.

In June of 1984, 3Com was investing in its future. Operations were moved to 1365 Shorebird Way in Mountain View, and expanded to fit the facility's 60,000 square feet. The new location provided much needed manufacturing space, and helped to house the onslaught of additional employees. 3Com made a conscious decision to refrain from building enclosed offices, the cubicles were here to stay (a nod to HP's egalitarian culture). The transition to the new site marked their first move with a significant revenue flow to maintain, but everything went extremely smoothly.[31] In August of that year, 3Com released its first annual report as a public company, documenting its 150 employees, and sales of $16.7 million. Key executives were being added to the team.

3Com's heightened public scrutiny made it more vulnerable from a legal standpoint. Mark Michael joined in 1984, just as 3Com was about to take off, and provided legal counsel to the company for nineteen years. His academic background included Stanford and UCLA. Mark became one of the longest running executive officers of the company, staying until 2003. Mark worked in the background—he wasn't someone you would see quoted in the press or pictured waving around routers and NIC cards. But he was one of the few who would experience first-hand all of 3Com's most notable leaders: Metcalfe, Krause, Benhamou, and Claflin.

Although Mark didn't come from a traditional technology background, he had been exposed to Silicon Valley technology starting at an early age. His mom was a rare female computer engineer, who helped install an IBM 1401 in

---

31. 3Com at this time had an OEM contract with Xerox for EtherSeries and a $1 million order from AT&T for transceivers, as well as OEM deals with HP, TI, and Zenith Data Systems, supplying everyone except IBM in the market for IBM-compatible computers. "Robertson, Colman, & Stephens, Market Research Report" Thomas E. Bliska, Jr. July 1984.

Stanford's Encina Hall, pulling all-nighters building metal coils and programming in assembly language, as Mark remembers. She moved to the Computation Center and worked on the Wilbur timesharing system (of "Orville" and "Wilbur"), and later worked for SRI's legendary Doug Engelbart at one of three nationwide networking centers. Mark was referred into 3Com by Howard Clowes, a partner at Ware and Freidenrich. With a reference from Doug Engelbart under his belt, which greatly impressed Bob Metcalfe, Howard Charney then offered the job to Mark. Bob was known to hate lawyers, but told Mark he was "OK." Having won his keep, Mark became employee number 168.

Chris Paisley, who became 3Com's CFO, joined in 1985 after working for HP and then Ridge Computers. Though Chris had been doing good work at Ridge and had the chance to help the company go public, he was ultimately convinced to move to 3Com, given the strength of their management team.

Bill Krause, who knew Chris from HP, had taken a huge chance in hiring him. Bill explained to Chris:

> There are three different profiles of people that I can recruit, I can go get a very, very senior CFO, who is on the downside of his or her career and this would be their last stop. Or I can get the very strong, solid, middle of their career CFO. But quite candidly, I think I'd have trouble attracting them to a company our size. Or I can get somebody that I hope is an up-and-coming CFO and that they'll grow into the job and do exceptionally well, and I put you in that category.

Chris responded to Bill, "Well, I want you to know, I'm going to give this my all. If it doesn't work out, it's not going to be due to lack of effort."[32]

Another key executive, Alan Kessler, landed at 3Com in 1985. Alan was a valued leader in marketing, sales, and customer service, and he eventually served as President and COO of Palm. His arrival at 3Com was preceded by significant exposure to emerging technology while working with Adam Bosworth[33] at Analytica, which was later bought by Borland.

In November of 1984, Bill issued an email which outlined a restructuring of the product divisions. The divisions were to be broken out in a move he believed would help scale the company towards $100 million in annual sales.

---

32. When I joined in 1989, I remember Chris saying that as soon as the company had sufficiently "made it," he was looking forward to a life of teaching, which was his first passion. Sure enough, shortly after stepping down later in 2000, and after turning down the Palm CFO role, Chris began teaching accounting at the Santa Clara University since 2001, where his classes are packed and in high demand.

33. Adam Bosworth and Brad Silverberg later headed up the Quattro project, moved to Microsoft, and led the MS Access project. As an early pioneer of XML technology, Adam has held leadership roles at BEA Systems, Google, and Amazon.

At the time of the memo, 3Com's annualized run rate was $50 million and it employed 252 people.[34] Howard Charney was put in charge of Network Adaptor Division (later changed to adapter), Larry Birenbaum ran Systems Software Division, Dan Robertson was VP of Network Servers Division, Chuck Kempton ran Sales and Marketing, and John Celii oversaw Finance.

In the fall of 1984, 3Com introduced and shipped the first of its internally designed network servers, called the 3Server. With a 40-megabyte hard disk, 30286 processor, Ethernet connections, and 3Com's own network operating system, the 3Server was designed as a network server from the ground up. Growth continued in 1985, with sales of $46 million, a whopping 278 percent increase over 1984. The production capacity of the 1365 Shorebird site was expanded, including the start of a second shift in manufacturing. For Fiscal Year (FY)[35] 1986, 3Com's simple mission was to lead in LANs for PC users, with clear objectives: make a profit, serve customers, achieve product leadership, and build a quality people organization.

The Convergence That Wasn't

In March of 1986, 3Com came within two days of merging with Convergent Technologies, which would have changed 3Com's trajectory and this story. This merger was intended to help add the "client" side of client-server computing. (Client-server technology is a model in which the computing power needed to perform certain tasks is distributed among a centralized server and group of client computers, typically the PCs on individual desktops.)

What Convergent brought to the table was, primarily, access to the market for workplaces that relied heavily on UNIX-based workstations. The weekend before the merger was called off by the investment bankers, there was a big "last 3Com" party at the legendary Keystone nightclub in Palo Alto, with the theme "Back to the Future."

Chris Paisley helped to explain why the merger unraveled at the last minute—essentially, Convergent was on the decline, while 3Com was rising. The 1-to-1 exchange ratio was beginning to look less favorable to 3Com. The investment bankers, Robertson, Coleman, and Stephens, were getting nervous. Chris felt he earned his stripes with Bob Metcalfe when he went to him and said, "Bob, I'm having strong misgivings about the deal. It's really awkward because I just arrived at 3Com. I'm the CFO. If the deal goes through, I'm no longer the CFO, so I'm worried that somebody would think I'm speaking out of self-interest, as opposed to interest for the company. I'd like your advice and counsel."

34. Internal email from Bill Krause to all employees. November 1, 1984.
35. 3Com's fiscal year ended in May, while Cisco's fiscal year ends in July.

Chris recalled how Metcalfe offered up his counsel, "He told me to speak up. But he also let me know that he was in favor of the deal and why he thought it should go through. But I did speak up." Chris recalled that Sandy Robertson was there, the named principal of the firm. He hadn't been involved at all previously. Chris explained:

> That alone sort of said there was going to be fireworks. They said they could not offer a fairness opinion. They didn't think it was in the shareholders' best interests. I remember asking them in the meeting, "What would be a fair exchange ratio?" It was at that point they gave an answer that I think did them in, in terms of their ever having future business with 3Com. They said, "We haven't been engaged to answer that question." They weren't getting much of a fee, to start with, because they were just supposed to rubber-stamp a deal that had already been negotiated and agreed to. So, they weren't exactly putting a large fee at risk, but they perhaps saw a chance to get a larger fee. While I think they were operating legitimately in the shareholders' best interest, trying to get a better deal for 3Com, the way they went about it sort of cut their own throat. There was an attempt to renegotiate that deal that went all night long as I remember staying up all night working on various stuff.

Paul Ely, the CEO of Convergent, later thanked Chris for his time and energy, and noted that his stance helped his reputation. Chris explained, "I think I also built credibility in that exercise with the board, because the board knew I was against the deal. As subsequent events unfolded and Convergent's fortunes fell and ours rose, it looked like the old story about sometimes the best deals are the ones you don't do. That was absolutely true in that case."

Convergent Technologies was eventually acquired by the mainframe-computer systems company Unisys. On a side note, Apple also tried to buy 3Com—twice. The first time was in 1983. It was initiated by Steve Jobs, according to Bill Krause. Bob and Bill talked to Mike Markkula, who was Apple's CEO at the time and Floyd Kvamme who was an EVP. 3Com rejected Apple's offer of $45 million, and went public the next year at close to a $100 million valuation.

The Convergent misfire triggered Apple's second overture. Bill recalled:

> Ten minutes after the press release saying the Convergent merger didn't go through hit the wire, my secretary came to the board meeting to let me know that John Sculley, Apple CEO at that time, was on the phone. John wanted to talk with us about merging. Bob

and I had dinner at John's house in Woodside with several other Apple executives and again we said that we were not interested.

Sticking to its Knitting (For Now)

After the dust settled on this failed merger, 3Com returned its focus to what it knew best: developing the technology that provided for networking computers to one another (for a while, at least).

With sales of $64 million, FY 1986 was another strong year (up 39 percent from FY 1985). 3Com now had 1,200 authorized resellers selling its products; the company sold products that connected approximately 100,000 PCs in 10,000 networks that year. Connections for AT&T, HP, IBM, and Xerox devices constituted 31 percent of the business, and there were co-marketing agreements in place with HP and DEC. Across all customers, the relatively new network server business amounted to 31 percent of sales that year, while adapters contributed 58 percent and software brought in the remaining 11 percent.

As business continued to grow rapidly, 3Com was again bursting at the seams. With a need for more space, the company moved to a site on Kifer Road in Santa Clara in December 1986, initially using two buildings with 160,000 square feet. Over the next few years, the company eventually occupied the rest of the site—four buildings total—with a total of 320,000 square feet.

The notion to move to Kifer Road was the idea of Abe Darwish, a newly hired, gregarious, and outgoing Facilities Director from ROLM, that later became 3Com's VP of Real Estate and Site Services. Abe explained how his family's journey as refugees from Egypt served as a role model for taking risks and joining companies like ROLM and later 3Com. Abe explained his risk taking instincts when joining 3Com:

> I stuck my neck out during the 3Com interview, and said, "We can do better on the impending move in Mountain View, I can move you guys prior to lease expiration on Shoreline and save us money," although I didn't have any concrete facts. But just my intuition told me that it could be done. Anyway, long story short, I got hired. Before I started, I did some research, and within a week of joining I made a presentation about the new location on Kifer, and fortunately for me and 3Com, we went on to create a space that fit great from not only a physical perspective only but also a cultural perspective.

The Kifer road site helped enable 3Com to grow its business and reinforce a company culture. 3Com employees and managers built their first world-class manufacturing facility. They found ample space for end-of-quarter parties in the courtyard, exhibited a sense of openness along with fun colors, yet provided quiet work areas, and shared its egalitarian values with its open cubicles—all these attributes supported the 3Com culture.

"Less" is More

One of the more unusual new products to emerge as part of this growth spurt had an interesting backstory. Bob Metcalfe led a small team with personnel from engineering, manufacturing, and marketing to bring out a low-cost "diskless PC." Diskless PCs were in some ways the precursors to today's Chromebooks, or stripped-down laptops that run apps and store data files in the cloud. Diskless PCs relied on connected file servers for storage (storage was very expensive) and print server functions.

The result—the 3Station diskless network station—was launched in May 1987. The device was managed through a separate division and business unit, and housed in off-site operations as a separate business unit. Although this line of business was dropped after a few years, many members of the 3Station team went on to other parts of 3Com and were key in new product and process innovations. Upwards of 100,000 3Stations were produced and were, at the time, innovative products. Larry Birenbaum explained how Bob Metcalfe had a different product name in mind for the 3Station—he wanted to call it the "Less." Compared to typical PCs, it was less a disc, less a floppy drive, less memory...and the list of what the "Less" was "less" went on. (Like many of Bob's ideas, that one was price-less.)

The Can-Do "Can't Say" Project

Rich Redelfs, an early marketing manager within the adapter division, explained how IBM, a key 3Com competitor, was also a key 3Com customer. Rich has a colorful background. He sang as a boy soprano in the top boys' choir in the U.S., and was even invited to sing in the Vatican in front of the Pope. This, he felt, helped with his confidence later in his career.

Rich explained that when he joined, IBM had spent $1 billion developing an engineering workstation computer based on emerging Reduced instruction Set Computing (RISC) technology called the RS/6000, and was new to this market. It also was using Ethernet, at the time when the rest of IBM was committed to the competing Token Ring approach. Inside 3Com, this IBM project was called "Can't Say." For example, if someone asked, "Rich, what are you working on?", his response would be "Can't say!" (IBM wasn't entirely anti-Ethernet. As Rich explained, for a little while 3Com also partnered on network management with IBM on "Heterogeneous LAN Management," which permitted Ethernet and Token Ring networks to be managed by a single platform.)

Rich went on to run OEM sales, taking over from Dave DePuy, who had helped 3Com in those important early 1980s as part of the team that got the first 3C100 transceivers and Ethernet adapters out the door. Rich noted that they eventually did away with OEM sales, and asked those partners to just buy standard products, with no further customization or private labeling for them (although later in 3Com's history in the 1990s, Dell and Gateway became OEM customers when the NIC chips were included on their motherboards).

By 1989, 3Com had shipped its first million adapters. It only took 18 months to ship the next one million adapters, and when Rich Redelfs left in 1999, they were selling four to five million units per month.

Chapter 6: The New Kid on the Block

3Com continued to enjoy sustained and impressive growth as a result of new product development, the expansion of customer markets for those products, and the rapidly increasing adoption of networking. But there was also pressure to revisit—after the ill-fated attempted merger with Convergent—efforts to achieve the kind of instant and dramatic growth that generally only comes from mergers and acquisitions.

In addition to generating growth overnight—at least as far as the balance sheets and revenue data are concerned—another impact of mergers is the sudden injection of fresh blood in leadership positions.

As part of the aftermath of 3Com's eventual merger with Bridge Communications, Eric Benhamou would wind up as 3Com's next CEO. That change in leadership ushered in a very different approach to corporate leadership, one that would result in a significant shifting of gears for the company.

Eric would lead the combined company as 3Com's CEO for a decade, from September 1990 until December 2000. (He then served as the board's chairman until the company was sold to HP in 2009.) In addition to Eric's role at 3Com, he also served as the CEO and Chairman of Palm for a time. In 1997, President Clinton appointed him to chair the President's Information Technology Advisory Committee. He was the recipient of the David Packard Civic Entrepreneur Award in 2007, and has been recognized by other U.S., French, and Israeli organizations.

Before diving into the events before, during, and after the 3Com-Bridge merger, it's helpful to understand a bit about Eric's personal and professional background.

Eric was born in a tiny village between Algeria and Morocco in the mid-1950s during harsh war times. His Jewish family was part of a mass exodus, and they picked France as their destination. In his early years, he found that France didn't offer the entrepreneurial culture he was seeking. As a child in Grenoble, he started a bike repair business. As Eric described, "I think I was eleven, this is when I wrote my first business plan. It was basically about building a moped repair shop. It occurred to me that there was a good opportunity there when I saw the kind of bicycles or clunky mopeds my friends were using, including myself. But the thrill was not so much that there was anything to invent. This was not a technology business. It was basically helping to organize a group of people around a common cause, sharing common profits, and doing things together that we could not do individually. That was what inspired me. And this is what has inspired me ever since."

After his schooling in France, he wanted to continue his education, and sought a master's degree in engineering from Stanford. Although he was interested in biomedical engineering, he landed in Silicon Valley in the mid-1970s. The microprocessor had just been invented and Xerox PARC had just started sending Ethernet packets between computers. The attraction of digital technology was impossible to resist.

After earning his second master's degree at Stanford, he opted to forego a PhD program, so that he could pursue his hunger for the entrepreneurial experience sooner.

His first gig was at Zilog,[36] the second microprocessor company in Silicon Valley (after Intel). By 1978, the company had built a network of sorts like the ones used today—with workstations, shared printers, and connected servers. Zilog Co-Founder Ralph Ungermann ultimately left, and founded Ungermann-Bass, and offered Eric the opportunity to join him at the new venture. But without his green card, Eric declined Ralph's offer. Eric wound up competing with UB; he joined Judy Estrin and Bill Carrico (who were married at the time) and became part of the early team at Bridge Communications—where his expertise with networks could combine with his entrepreneurial passion.

By 1987, there were a handful of significant networking companies in the world—Network Systems, 3Com, Bridge, Ungermann-Bass, InterLan, and Cabletron. (UB would be acquired by Tandem Computers nine years later for $260 million.)

A Merger on the Rebound

Eric had recognized that a merger of a LAN-centric company like 3Com with an internetworking-centric company like Bridge (whose products helped

---

36. Zilog was co-founded by Federico Faggin, an Italian physicist, inventor, and entrepreneur, widely known for designing the first commercial microprocessor. He led the design group during the first five years of Intel's microprocessor effort. Faggin created in 1968, while working at Fairchild Semiconductor, the self-aligned MOS silicon gate technology (SGT) that made possible dynamic memories, non-volatile memories, CCD image sensors, and the microprocessor. Ralph Ungermann was a co-founder.

customers create and manage "networks of networks") would come with risks, but also saw that there were risks involved in not pursuing a merger. "I would say it was a defensive merger with 3Com. Defensive because we felt that inevitably, we were going to bump against much, much larger companies like IBM, Digital, HP, and many others. We felt we needed critical mass, which we could gain through a merger with 3Com. This would give us the staying power to grow and rival the marketing and sales muscle of much larger companies."

Around this time, John Hart, a VP of Engineering for Vitalink Communications, had successfully found a way to take Bridge products and use them with their satellite connections, and built an enormously successful business with DEC. John tells the story of what happened after collecting a $15 million insurance claim following the explosion of one of their satellites:

> Tom Perkins [of the VC firm Kleiner Perkins] says let's just divide up the money and go home. I said we've got this one project with DEC, that might be really interesting. Perkins let us keep it alive. We ended up going to DECWorld ... Fred Baker [a Cisco fellow for the last 22 years] helped me put up a display of our networking DEC's Massachusetts lab on-line—this silenced the entire DECWorld, who [were] able to see it on their monitors on the conference floor.

This was the birth of the explosion in remote bridging, and put Vitalink on the map. They became the first independent networking company to go public.[37]

Bridge had become entrenched in its focus on networking mainframes just as the market for PCs was beginning to take off but also had begun making routing devices (a market that Cisco would later dominate). Recognizing the shift in the marketplace, and realizing the difficulty of catching up to the lead of its PC-focused competitors in those high-growth markets, Bridge was looking to expand its market. Merging with 3Com would make a lot of sense—at least in theory.

Bridge had earlier raised $1.8 million with Weiss, Peck, and Greer, the precursor to what today is venture capital firm Lightspeed Partners. By the end of 1982, Bridge had built their first beta product for UCLA, networking the school's PDP 11 and VAX machines. Bridge's products were well-suited for several high-growth markets, such as the increasingly popular Unix computers that used the TCP/IP model for networking. In 1985, with $35 million in annual revenues, Bridge went public. Eric quietly ran engineering and kept his focus on internal matters. Bridge's growth continued; by 1987, they were generating $50 million in annual revenues.

---

37. Vitalink deserves its own story. For more history, visit our website at 3comstory.com/tech.

3Com was the bigger company—about 50 percent larger than Bridge—but 3Com was also feeling the need to remain competitive in a pond that was rapidly becoming too big for small fish. In light of the failed merger with Convergent, 3Com was ready for another opportunity to ensure it remained on a fast-track for growth. Bridge became part of 3Com, in exchange for $200 million in stock.

3Com was now one of the leading companies focused on the exploding local area network (LAN) business. The company had established the standard in Ethernet adapters for personal computers, while gaining experience in the LAN-systems business. They had built and maintained a powerful retail/distributor channel. And Eric would ultimately become 3Com's CEO.

But there would be a challenging period of adjustment before that would happen. As Cisco's long-time CEO John Chambers once observed rather diplomatically, "In a merger of equals, you stand a very good chance of stalling out both companies… It can be a major distraction to both."[38]

Water Over Troubled Bridges

As is often the case after a merger, there wasn't a clear roadmap for what should happen next. There was uncertainty and tension, largely rooted in the existential question: what are we now? Was the new combined company a "computer company" (focused on networking products for computers) or a "networking company" (focused on communication solutions connecting computers and networks of computers typically sold at the enterprise level)? As Eric said, "There were fundamental issues that had not been resolved. We were going to figure it out after the merger was done."

As part of the merger agreement, Bill Carrico became President and COO of 3Com and Bill Krause became Chairman and CEO. Together, they formed an "Office of the President," which, as Judy Estrin noted, "is one of the things that created the problem at the top." Bill Carrico was an impatient leader, and very bright. He had no tolerance for people who were slow to act, or who favored process over substance. And his relationship with Bill Krause after the merger was tense. Dick Bush, Bridge's first VP of Sales, recalled that "Bill [Carrico] would pick up a report he didn't like and throw it back at you." Bill Krause, too, is remembered for his own fiery expressions from time to time. The stylistic differences of everyone involved made finding resolutions to the issues that arose after the merger much more complicated.

With his experience in computers, Bill Krause saw networks as a solution for connecting computers, using technology that ultimately revolved around computers and data storage devices. There were, as Jerry Dusa remembers,

---

38. Paulson, Ed. "Inside Cisco: The Real Story of Supercharged M&A Growth". Wiley. 2001.

early indications that there was interest in the vision of network of networks. But, at the same time there was some realization on the part of the board, Bill Krause, Bob Metcalfe, and others that there would be a great market in selling PC-based LAN boards, servers, and workstations. If the PC market was booming—and if LAN boards were soon going to become part of virtually every PC sold—then it was vital to stay focused on this seemingly golden opportunity to sell 3Com components and products for PCs in addition to his view of client-server computing, in similar fashion to Sun Microsystems and Apollo Computers, the leading workstation makers at the time.

This realization ultimately fueled the decision for 3Com to remain focused on the PC and client-server computing space and allow the nascent market for networking networks together to take a back seat. Bridge had been more focused on internetworking and the related connectivity and communication-server products, so this left the acquired Bridge employees, including its leadership, wondering where and how they fit in.

There were also some significant cultural differences between the employees in the original 3Com camp and those that were assimilated from the Bridge camp. 3Com's culture borrowed heavily from HP. The company was process- and detail-oriented, inviting debate, seeking consensus. Bridge, by contrast, had a more entrepreneurial culture, and followed a top-down management style under Bill Carrico. When recalling the merger, Bridge Co-Founder Judy Estrin acknowledged that, at the time, each company had identified something in the other that it needed. While the merger made strategic sense, Judy noted that, "Because we knew each other well, we did not adequately understand the differences in each other's visions and styles before the merger." Judy felt that these fundamental differences in each company's vision were not adequately explored before merging. At the time, the two companies were literally next door to each other in Mountain View, where Google houses its Google X initiative today.

In Ed Paulson's 2001 book *Inside Cisco*, Bob Metcalfe recounted, "3Com was two or three times bigger than Bridge at that time, but we treated it as a merger of equals, which was stupid. We ended up with two heads of sales, two heads of France, two heads of Germany, two heads of marketing, two heads of engineering, and they spent the next couple of years trying to kill each other."

Each company also had very different concepts of who their customer was and how they managed their relationships with them. Bridge had been driven by the needs of their direct enterprise customers, while 3Com held their customers at a distance, selling instead through distributors, resellers, and OEM channels. Judy found it strange that some of the management at 3Com "didn't even know who their customers were." Not only were there fundamental differences in their sales and management styles, but there were significant differences in company philosophies and their approach to conducting business. Perhaps the biggest gulf between the two halves of the merged company stemmed from the recognition that each side harbored a technological vision for the company that simply weren't compatible together.

This issue of direct vs. channel sales methods was less of a problem in Europe. As the 3Com and Bridge folks merged, they had already been working hard to penetrate the direct enterprise customers with 3+Open (the first LAN operating system developed with Microsoft), client-server computing, and bridging products. But the method of selling in the U.S. was largely determined by 3Com and Cisco's early product DNA. As the PC revolution launched, 3Com had early success driving a need for mass market products while using a distributor/integrator channel strategy. Cisco, entering a few years after, benefited from their core networking products sold by an impressive direct salesforce, though they later adopted a distribution channel strategy as well.

As one executive reminisced, "3Com was much bigger than Cisco internationally for some time, and we had good presence in accounts, some of it working directly with end users, sometimes working in conjunction with integrators. But you can't win those [sorts] of wars if you don't really get in the U.S. core market." Ultimately, Eric's vision of creating a networking infrastructure company would win over the support of the board of directors. But not without an awful lot of water under the bridge.

Walls and Bridges

In 1987, Ronald Reagan gave his memorable speech at the Berlin Wall during which he implored his Soviet Union counterpart Mikhail Gorbachev to "tear down this wall." That became a reality two and half years later. Depending on who you ask, the easing of tensions between 3Com and Bridge either happened a little more quickly—or took an awful lot longer.

One of the reasons behind the ongoing tensions was that not everybody agreed exactly what business 3Com was really in—computers, networking, enterprise systems? Today, there are pretty clear boundaries between those markets, but in the early days of 3Com and its competitors, the borders weren't always clear.

As Eric observed, "Bill Krause and Bob Metcalfe felt 3Com was building a replication of Sun—their focus was on workstations, servers, and software. Bill Krause had a computer-systems background, I had more of a networking background and so did Judy, and we wanted to build a networking infrastructure company." While Sun was anchored in the UNIX operating system, 3Com would (for a while) be tied to the Microsoft-IBM first-generation multitasking operating system, known as OS/2, along with its networking extension, LAN Manager (or under 3Com's brand, 3+Open).

And 3Com and other networking companies (Cabletron, SynOptics, Ungermann-Bass, etc.) weren't only in competition with each other, but they had to fend off companies far larger than themselves—like IBM, DEC, Wang, and Intel—that were starting to play in the networking arena as part of their broader product offerings. Smaller companies began to sprout up, like Cisco, Wellfleet, Vitalink, and others.

Reminiscing about the Bridge merger and its aftermath, Eric recalled:

> I felt our approach had a very different feel than building computers. Clearly there were some smart networking people at Sun, otherwise they could not have built the company they built. But we felt networking was still a very distinct discipline and focus. The way the industry evolved, the networking industry started off as an outgrowth of the computer industry, but eventually it became completely its own separate and huge critical mass. Obviously, a lot of it was dominated by Cisco—but for a long time we were neck-and-neck rivals with Cisco in terms of revenues, growth rates, market cap, and so on. Networking was a very different discipline and eventually we took that path—but for the first couple years post-merger, there was a lot of strategic ambiguity and, eventually, the board had to intervene. Not just in terms of choosing a leader for the next few years, the next decade, but to bring some clarity to the strategic direction.

The absence of this early strategic focus would hurt 3Com from a sales perspective, and it would also trigger a "brain drain" of Bridge's staff. In *Inside Cisco*, Ed Paulson emphasizes the double-whammy impact that a company experiences when it fails to execute an acquisition effectively, "Notice that there is an implicit threat associated with not handling an acquisition within the same industry properly. Acquired people who are not happy with being acquired will leave to work for other companies, typically within the buyer's industry. These people now become competitors, when a well-done acquisition could have kept them in the buyer's organization, working for the buyer, not against it. Experience also has shown that the most qualified people are often the first to leave since they have the easiest time obtaining new employment. So not only does the buyer lose the most qualified, key acquired people, but most likely the buyer also loses them to a competitor. This is a double injury to the buyer."[39]

---

39. *Inside Cisco*. Pg. 141.

While the merging of 3Com and Bridge appeared to be a great pairing of technologies on paper, it ultimately did not pan out as expected. Krause's view of focusing on client-server computing, conflicted with Bill Carrico's and Judy's view of going after internetworking. Nevertheless, some at 3Com attempted to put Bill Carrico in as COO and President of the combined company briefly. The marriage wasn't meant to be and after nine turbulent months at 3Com, in May of 1988, Bill and Judy left the company.

In retrospect, Judy feels that stepping away was the right decision at the time, given that a warring difference in visions could hurt the company. She believed, "It is not healthy for a company to have long-term friction at the top, you need clear leadership." More than half, and many of the best, of the Bridge folks followed suit, except for Eric. Some Bridge folks found their way to SynOptics Communications (SynOptics), but most of them went to Cisco, the company that would increasingly be 3Com's most challenging rival. Cisco's CEO at the time John Morgridge, who had joined Cisco in 1988 as Judy and Bill were leaving 3Com, later told Bill Messer, 3Com's first Asia Pacific Regional Sales Director, that the talent they acquired as a result of 3Com's deal with Bridge was amazing, simply the best thing that could have happened to the company. In 1988, Cisco had 34 employees, while 3Com had 1,348 employees.

Cisco went on to accelerate its router development with the help of defecting Bridge employees, while Krause stayed the course with client-server computing, not fully investing in the products that Bridge had brought to the table. This left 3Com with a hollowed out strategy for the router market, and let Cisco capture a dominant share of the enterprise space. As Judy would acknowledge, Cisco took advantage of an opening left by 3Com. "In everything in life," she said, "a lot of it is skill, and a lot of it is timing, and a lot of it is taking advantage of opportunities that are created by someone else."

This strategic divergence, whereby 3Com spent the next three years on client-server computing and adapter cards, while Cisco, along with their Bridge defectors pursued and developed their knowledge around routing and internetworking, would turn out to be perhaps the most important turning point in 3Com's history. In this period, Cisco was getting cozy with the largest, most complex enterprise accounts, building a "network management" story and a strategic incumbent advantage, while 3Com honed its distribution channels, selling less complex, higher volume, lower priced products in mass. Cisco's VP of Marketing in the early 1990s, Cate Muether, who had moved over from 3Com after the Bridge merger, said of her time at Cisco, "The orders would just come in on the fax—it made marketing so easy."

Eddie Reynolds, a management and HR consultant, was hired by Debra around the time of the Bridge acquisition, and his task was to bring structure around organizational development and leadership. His first project was Bridge assimilation. He later helped run all the ExecCom, OpCom, and 3Com At the Half strategy events for most of 10 years—under Bill Krause first, and then Eric Benhamou. 3Com At the Half was the annual management meeting that grew larger each year, where discussions were held about strategy, vision, values, and so forth. ExecCom was Eric's direct reports of initially six executives, but later grew to over fifteen by the end of his tenure in 2000. OpCom included the direct reports to ExecCom members.

A New Competitive Landscape

Sales continued to grow at such a rapid clip that it was driving changes in every corner of the company. The most obvious change was that they needed to evolve the sales operation. Sales in 1986 topped out around $64 million; in 1987, sales had almost doubled to $110 million. Jerry Dusa was the U.S. sales manager during this time. He only stayed at 3Com for a few years, but he remembered the time well, "My tenure there was punctuated by high growth rates, real strong market acceptance, and the forever challenge of trying to match supply and demand of our product during a time of tremendous growth."

When he arrived in 1987, the U.S. salesforce was made up largely of direct sales representatives. Alongside them were a few people focused on OEM sales, working to get board-level products sold to PC manufacturers rather than as some aftermarket product customers would buy and install themselves. As time progressed, they added distributors to sell to large end-users, such as Fortune 100 corporations, in addition to the so-called two-tier distributors, championed by Ralph Godfrey, SVP of Sales who was recruited from HP in 1990.

The distributors sold to value-added resellers (VARs, or companies that acted as middlemen selling customized systems to larger customers). On top of this, they grew their OEM channel serving manufacturers. Jerry Dusa said, "The sale of LAN products for PCs was skyrocketing, and we needed as many ways to get products to people as you could possibly find."

To manage the growth in sales, 3Com was also experiencing a burst in personnel. As Dusa put it, "We were adding people by the handful." Growth in sales was happening so fast that Dusa recalled it as a blur. In fact, he was too busy to pay attention to what was going on simultaneously with the Bridge merger. He and his salesforce were "110 percent over-extended every single day, all day long, trying to keep our noses above water."

With the benefit of hindsight, however, the most significant transformation driven by the explosion in the networking market was probably the shift taking place in the competitive landscape. As the team embarked on their FY 1988 plans, the company in general and Bill Krause in particular held the following view about who their most significant competitors were (prior to the Bridge merger):

- Computer companies: IBM, along with DEC, HP, AT&T, Wang Laboratories, and Data General.
- Workgroup companies: Sun, as well as Apple, Compaq, and Apollo.
- PC LAN companies: Novell, plus Banyan, Western Digital, and AST.
- PC software companies: Novell again, and Microsoft, Lotus, and Ashton-Tate.
- LAN systems companies: Ungermann-Bass, plus Bridge Communications, Sytek, and Interlan.

A few things are apparent from this list. First, 3Com (like many tech companies then and now) had customers and partners that doubled as competitors. It was also clear that 3Com was heavily focused on the market for enterprise work-groups, and less worried about what would become the Internet of today, with its tremendous demands on wide-area networks (WANs) and internetworking that now support the World Wide Web. While the Bridge acquisition, on paper, addressed that emerging market, 3Com's resources, sales strategy, and execution did not follow down Bridge's path. Finally—and most significantly—it's somewhat shocking (at least with the benefit of hindsight) that the budding 800-pound gorilla in the room (Cisco) hadn't even appeared yet on that list of competitors.

At least Bill Krause and company had bought Bridge—what if they hadn't? 3Com's flame may have expired sooner, as many networking companies did while 3Com remained standing. And while no one would have seen this coming, in the decades down the road, internetworking systems products would turn out to be the ultimate endgame for 3Com, albeit from a joint venture in Asia.

But at least one outside observer—none other than Steve Jobs—had a sense of where the market was headed, and where it wasn't. Several executives recalled that Jobs, at an executive offsite meeting that Bob Metcalfe had invited him to join, had importantly asked, "Why would you want to be like Sun? or Compaq? Do you really think you can buy in scale the way Compaq does for servers? Why aren't you focused exclusively on networking?"[40]

3Com was, however, pursuing what it viewed as a major next chapter in digital technology. Around this time, Bill Krause had partnered with Microsoft on the next-generation operating system it was creating with IBM. 3Com's role on the OS/2 project was to build computers, servers, and workstations that

---

40. Way down the road of his career, when Eric had completed his CEO tenure in 2000 and NeXT had been sold to Apple, Steve Jobs in his interim CEO role at Apple attempted to hire Eric as CEO. While Eric was and is a big Apple products fan, he had no interest or inclination to consider the job seriously.

would run on OS/2 software. All of this would then be branded as under the umbrella of 3+Open. OS/2 software was designed to replace the ubiquitous but antiquated DOS operating system. One goal was to compete with Novell to drive its network operating system (NetWare) out of business; another was to become more like Sun by building computers and workstations that leveraged innovative in-house software. OS/2 did not succeed, as Microsoft eventually abandoned it in favor of Windows 3.1 and Windows 95, which incorporated many of the ideas first showcased at Doug Engelbart's landmark "mother of all demos"[41] much earlier in 1968.

Michael Smith, a Bridge Communications employee who later became a 3Com product marketing manager, also noted that the terms of the deal with Microsoft gave Microsoft all of the rights to 3+Open after version 2. In essence, Bill was investing the profits from the adapter division into something which, fairly soon, 3Com would no longer have any ownership of or control over. It seemed apparent that Microsoft would simply incorporate the underlying technology into their operating system. As development progressed, Michael said, "The Redmond guys would come down, and you could tell they were really, really brilliant people, and they would make very ignorant assumptions, and they'd ask questions that indicated their naiveté, and I think it would kind of annoy some of the people at 3Com, because they understood it cold.

But what I noticed at every meeting was they'd come in, they'd ask a bunch of questions, and the 3Com engineers would just say the answer, and then they'd leave. They didn't ask the same question twice. And I realized, they're just sucking it up. And eventually they incorporated everything into the operating system, which effectively made the NICs a commodity."

The paranoia about Microsoft absorbing the networking technology into its own products was well-founded, as Microsoft ultimately absorbed the networking software 3Com had worked so hard on into Microsoft's own operating system, and they separated ways from IBM. IBM drove the project on its own for a few more years until OS/2 faded into obscurity. 3Com had also invested heavily in something it thought would blossom into a huge market, which instead became commoditized and invisible inside the Microsoft software.

However, it would take a bit longer for that outcome to sink in. As 3Com ended its first decade on May 31, 1989, the mood of the company was upbeat. The company employed 1,922 people. Sales were up 53 percent—to a whopping $386 million—from the year before. There was the goal of reaching 3Com's "first $100 million quarter" within the next few quarters. Net income was $34 million, or $1.18 a share, a 52 percent increase over the prior year. Somewhere during

41. To read more about "The Mother of all Demos" visit 3comstory.com/tech.

the year, the company shipped its one millionth adapter. Still, FY 1990 sales would plateau at $419 million, only 75 percent of 3Com's target of $558 million, one of the rare misses in Bill Krause's disciplined planning and goal-driven company.

In addition to selling 3+Open LAN Manager, the network operating system based on the Microsoft LAN Manager, 1989 also saw the company's announcement of NetBuilder I, the foundation for next-generation internetworking products such as NetBuilder IB/2000 local Ethernet routing bridge.

Mark Michael shared an interesting and colorful story about 3+Open and the Chinese marketplace:

> At one point early on in the Chinese push for advanced technology in the mid-1980s, 3Com accounted for five percent of all computer sales in China. How? The Chinese were placing a big bet on PC networks and were happily misappropriating our 3+ operating system, which the DoD refused to issue permits for us to export lawfully. It was the mid 1980's. Richard Perl was Assistant Secretary of Defense and pursued the Reagan ideology of communist Evil Empires. LANs that allowed printer sharing, file serving, and email were deemed to be weapons.
>
> We tried for almost a year to get an export license. The Chinese government sent over a delegation and we later met with them in Beijing. The leader of the group was Mao Zedong's niece. We negotiated the draft of a proposed joint venture, subject to U.S. export licenses being issued. These never happened because of the absurd classification of a 3Com LAN as a command and control system for weapons. The Chinese reverse-engineered the 3+Open software and created their own Chinese language version, creating huge demand for 3Com's hardware that was sold into the Chinese market at favorable prices. Best thing about this situation was the absence of any warranty or support obligations on the Chinese bootleg software.

Chapter 7: Rebooting Our Bootstraps

Another major product launch that took several years to come to fruition was the Etherlink III. After the merger with Bridge, Bill was especially interested in transitioning 3Com away from hardware in favor of software. But hardware products (largely network adapters) were still 3Com's bread and butter, generating most of the company's revenue.

Hardware Products Division. Offsite Meeting, Jan 13, 1988. *Source: Dan Robertson.* 1: Andy Verhalen, 2: Rick Wilson, 3: Alan Jewett, 4: Pitts Jarvis, 5: Jim Hayes, 6: Jim Schelcher, 7: Larry Nunes, 8: Reinier Tuinzing, 9: Nat Tinkler, 10: Howard Charney, 11: Sandy Seppala, 12: John Nauman, 13: Gordon Smith, 14: Ron Crane, 15: Hon Won, 16: Gerald Petak, 17: Jeff Harry, 18: Claude Ezran, 19: David Schwartz, 20: Dan Robertson, 21: Niles Strohl, 22: Richard Lee, 23: unknown, 24: Jack Moses, 25: unknown 26: Mark Hoke, 27: Mary Stinar Lenehan, 28: John Housman, 29: Eric Larson, 30: Gary Rosekind, 31: Dennis Caravalho, 32: Gary Wood, 33: Perry Gluckman (management consultant), 34: Mike Nunez, 35: Bob Lasswell, 36: Stan Dutrow, 37: Chuck Leis, 38: Mike Lawrence

Andy Verhalen, a key hire recruited from Intel, who ran the Transmission Systems Division (which included adapters, 3Server hardware, and 3Stations) from 1986 to 1991, noted that cutting the earlier deal with Microsoft on what eventually became 3+Open was "crazy" even at the time (1988). "The company started as a pure hardware company building transceivers and network adapters. They tried to become a software company. It frankly wasn't where the DNA of the company was."

He saw a lose-lose situation but lacked the political power and influence to intervene. He recalled how impossible it would be for 3Com to compete with Novell in the network operating system business, given that Novell could just make their product "lean and mean" while 3Com would be restricted by much of Microsoft's existing software. "We have half our hands tied behind our back...If we're successful, Microsoft is just going to take it away from us." Strategically, it made no sense for 3Com to do all this development work for Microsoft when, at the end of the day, "They took it away and ended up owning it." The worst part, Andy explained, is, "We could have been investing in routers and bridges, and network management and infrastructure ... that's where our DNA should have taken us. We could have owned the plumbing and let Microsoft own the client ... and Cisco would have never existed."

Andy initially saw another opportunity to compete with SynOptics in the market for structured wiring products, while simultaneously leveraging products from Bridge's catalog. (SynOptics became best known for making it easy to create networks using the existing and ubiquitous twisted-pair telephone cables, and creating a successful hub product based on IBM's popular Token Ring for Ethernet.)

While Andy approached Bill Krause in 1989 to consider this hub strategy instead of focusing on adapters, Andy found that Bill did not want to spend any money on hardware. He pitched an idea to build a new hub that would not only serve as the backbone of 3Com's transmissions systems, but would also be able to integrate Bridge's products, which 3Com had yet to adequately take advantage of. Recalling his frustrations many years later, Andy remembers trying to convince a reluctant Bill, saying "Let's go compete for this business. It's the plumbing of the network and it really plays to our strengths. And by the way, it's going to become a huge business." Network penetration, he argued, was only 10 to 20 percent, and would be growing to 100 percent. The exploding growth in the PC market would also help fuel the market for a hub product.

Despite Andy's valiant efforts, Bill remained unconvinced. Bob Metcalfe told Andy that he needed to make it clear that this could be a cash cow big enough to fund Bill's broader strategic goals. And that got some attention: Bill gave Andy permission to spend 5 percent of his revenues on R&D—although that still wasn't enough to fund the hub project.

Instead, they decided to develop new adapter technology, 10BASE-T, using twisted-pair cabling as well as their own integrated circuits. By 1989, 3Com had lots of competitors in the Ethernet adapter market, so this would help them to strengthen their differentiation. Looking for a game-changing product, Andy asked his team, "All right, how do we improve performance and lower costs?" Unfortunately, he kept getting the same unsatisfactory answer: the product ideas all required copious amounts of expensive memory. "We were stuck."

Andy Verhalen had come from Intel and recognized that if we really wanted to make money, we needed to "own our own silicon" (that is, make our own chips that were tailored specifically to our products).

Andy was focused on two ASIC projects (application-specific integrated circuits customized for a specific task). On the one hand, if cost wasn't a factor, how could you build the most powerful adapter possible? And on the other hand, how much power could you pack into a chip if you needed to keep costs low? Ultimately, the low-cost Etherlink chip eventually became the real profit-making engine for 3Com. Andy's very carefully thought out product development and ASIC decisions probably made more money for 3Com than anything else the company ever did.

Looking for answers outside of his team, Andy pulled Paul Sherer, a brilliant engineer, over from the software division, where Paul was frustrated with his own position. Paul had graduated from the University of Alabama, worked with the U.S. Army and Air Force as an engineer, then joined Masstor Systems for two years, before joining 3Com in 1984. Recognizing Paul's exemplary work on the 3Server, Andy challenged Paul to "figure out how to change the game." Ultimately, Paul did just that, developing adapters that quickly exploded their market share and revenues during and after Andy's oversight. FY 1987 adapter sales were $65 million (out of $110 million in total sales, or 59 percent of 3Com sales). By FY 1996, 3Com reached $950 million in adapter sales (out of $2.327 billion in total sales), an increase of 1,461 percent in less than ten years.

Paul Sherer's solution is an inspiring story. As Andy Verhalen recalls, "He re-architected it from scratch." Paul's design process took him back to the basic principles of the Ethernet Blue Book[42], the "bible" that defined the standard, and figured out "bit-by-bit, packet-by-packet, how does [Ethernet] work?" Paul was not concerned with the predominant Ethernet products at the time, either 3Com's or those of its competitors. An innovative solution would require him to solve a problem that no one had tackled before.

Paul figured out what many could not see: that the main factor limiting Ethernet performance wasn't throughput (how many bits you could push through in a given amount of time), but latency (how much the signal lagged when you pushed it). As Andy explained, bigger buffer memories were being added to Ethernet products to improve throughput. But when more time was spent buffering, the latency increased.

Paul set out to eliminate the need for buffering by making the connection between the client and the server as fast as possible. By building buffering

42. In 1980 the DIX (DEC, Intel, Xerox) "blue book" Ethernet specification was published. It was the basis for the development of IEEE 802.3, the Ethernet standard published in 1985.

into the adapter's silicon, Paul ended up designing a key component of one of 3Com's most successful hardware products, the Etherlink III, which utilized Paul's new architecture in a very cost-effective form. The Etherlink III also boasted a new transceiver design by engineer J.R. Rivers that was able to improve performance while being manufactured at a very low cost.

Prior to Etherlink III, packets were transmitted, and when complete, the next packet would then go. But with Etherlink III, it had everything lined up so that you could start transmitting the first part of the packets, and before you finished, you had already lined up the next packet and the next packet and the next packet. Everything was "pipelined" in this manner, or what was branded as Parallel Tasking. In microprocessing terms, you fetched the next instruction while you're processing the current instruction. The Parallel Tasking brand was powerful and was included in the product introduction.

Etherlink III officially launched in mid-1992 and took off like crazy, according to all involved. Doug Spreng was being handed the baton from Andy Verhalen around this time. Etherlink III had 21 percent market share when Doug arrived, and eventually won 3Com a 50 percent share of the market for network adapters, It was one of the key factors driving up 3Com's stock price over the next few years. Perhaps bitter sweet, as Eric was hoping the systems products would drive stock price. As Andy put it, "It wasn't success in network systems. It was success in our original business, network adapters and transmission systems." 3Com had reinvested in its roots—in its own history—and it paid off handsomely. Andy Verhalen's vision and implementation, along with brilliant engineering, ushered in an era of adapter card domination with the Etherlink III; Andy's contributions cannot be overstated.

The Etherlink III may have also helped lead market acceptance of Ethernet at the expense of IBM's competing Token Ring network communications protocol. Up until this period, it was a struggle competing with IBM. 3Com was the dominant Ethernet adapter player, but there were hundreds of companies competing in that space—and it wasn't clear that it would prevail over Token Ring.

The Ethernet vs. Token Ring Horse Race

IBM had established an early lead in adapters in the mid-1980s with a technology known as Token Ring. They had implemented it by repurposing standard telephone cables for data communications, taking advantage of the low-cost and prevalence of unshielded twisted-pair (UTP) wiring, which would become dominant by the early 1990s. But Ethernet's popularity grew throughout the 1980s, helped by the arrival of Thinnet coaxial cables, which were less bulky and easier to install and connect. In the late '80s, AT&T introduced StarLAN, a flavor of Ethernet that operated on UTP phone lines, in a hub-based star network topology, which offered the same advantages for network management and troubleshooting that Token Ring had. SynOptics launched a version of Ethernet that would run on both fiber optic cable and the twisted-pair wire that IBM used, based on technology Ron Schmidt had developed at PARC.

Ultimately, Schmidt commercialized the cabling system he developed at PARC that supported both Ethernet and Token Ring by co-founding SynOptics as a separate company that Xerox invested in. LattisNet, as it was called, was a hub-based Ethernet version supporting both fiber optic cable and the IBM cabling system, and later evolved to embrace the 10BASE-T protocol in 1987.

Although there was some internal frustration waiting for Etherlink III to get out the door, once it shipped it provided a compelling competitive edge against other adapters and against the competing Token Ring approach. Token Ring began to lag in the marketplace, partially due to technical limitations, but more importantly, because it "lacked the socioeconomic force that generated the price reductions and innovations that secured Ethernet's long-term competitiveness." In other words, it did not achieve the widespread adoption and volume that would have driven economies of scale, enabling the cost economies that Ethernet enjoyed.[43]

There also was skepticism about the interoperability of IBM's own products. In order to grow its market share and keep prices and margins high, IBM required third-party Token Ring products to use IBM-compatible chips made by Texas Instruments. These turned out to be inferior to IBM's own chips that the company used in its own products. On the one hand, this gave IBM an edge against competitors, but it also created a disincentive for other companies to embrace and expand the market for Token Ring products. While the Ethernet ecosystem was encouraging the development of a diverse array of products, there was a far less dynamic entrepreneurial community revolving around Token Ring.[44] Some interesting statistics:

- 1989—2.2 million Ethernet adapters, vs. 1.4 million for Token Ring.
- 1995—23.7 million Ethernet adapters, vs. 3.8 million for Token Ring.

Ethernet's edge had moved from about a 50 percent lead in 1989 to a six-fold advantage by 1995.[45]

---

43. von Burg, Urs. "The Triumph of Ethernet: Technological Communities and the Battle for the LAN Standard". Stanford University Press. 2002. Pg. 187.
44. "The Triumph of Ethernet" Chapter 7.
45. "The Triumph of Ethernet" Pg. 193.

THE WANDER YEARS
(1990–1997)

The second decade, give or take a few years, is argued by some to be the most important in setting the trajectory of 3Com's path for the rest of its existence. Earlier, 3Com's acquisition of Bridge Communications in 1987 made sense as a strategy to keep the company at the leading edge of networking. But Bill Krause and Bob Metcalfe had decided to make internetworking the secondary priority to their vision of transforming computing. The company spent those three years post-Bridge merger squandering[46] the profits from adapter sales on developing software with Microsoft and other computing platforms, such as diskless workstations and 3Servers. Cisco gained a three-year lead from 3Com's stalled entry into this evolving market, and even lured much of Bridge's top talent away from 3Com.

Ultimately, 3Com arrived at their first big pivot by 1990—to focus the company on solutions to create networks of networks. As Eric Benhamou took over, he focused on promulgating a new vision for the company, one that would include a great deal of invention as well as aggressive acquisitions. The new and powerful vision—dubbed global data networking (GDN)—was coupled with Eric's thoughtful, analytic, and more consensus-driven (some might say more passive) approach to decision-making and he helped 3Com become the top player alongside Cisco in the 1990s. 3Com also set the pace early with its acquisition strategy, which Cisco quickly adopted and aggressively pursued.

Chapter 8: Shaking—and Shaping—Things Up

In addition to these shifts in the 3Com product mix, there was a growing recognition of the need to evolve 3Com's overall sales and marketing strategies, as IT directors were increasingly demanding a direct relationship with their suppliers. Brad Mandell, a sales executive, arrived at 3Com in 1989 when the San Francisco 49ers won the Super Bowl, but had to miss the game while driving to a sales meeting in Tahoe. Brad felt that sales was already missing the enterprise angle, "It was interesting that 3Com never went all in to attack the enterprise. We had great products, great solutions, and we were just completely outmanned by Cisco. Part of the reason was because we were a channel-driven company because of the adapters. That was our sales DNA when I arrived. Adapter fulfillment, versus building enterprise solutions, and account selling, it's just very different. We were 'one arm tied behind our back' because of the marketing and sales machine Cisco had."

---

46. At least one executive weighed in that "squandering" is a bit harsh; perhaps it was innovative risk taking and perhaps just ahead of its time. There is no doubt there was a large and robust market in these areas, but there were much bigger and more established players (such as Compaq) in those markets. Our DNA was founded in networking, not computing.

Internationally, 3Com was quite popular. One executive noted:

> Our customers in general really wanted us to be there. Although Cisco dominated many of those accounts, as you can imagine, enterprise customers don't like to have to choose just from one supplier. You end up with pretty bad pricing, hence why Cisco had fantastic gross margins, they didn't really have anybody competing hard with them in the enterprise. So, we did have good enterprise prospects who were willing to bet on us. They did think 3Com was an accomplished technology company. They thought our product quality brand was really highly regarded. And they wanted us to bring something into that space.

The company's board of directors felt that two changes were needed: a change in leadership, and a smarter and better articulated strategic plan.

Eric Benhamou, along with sales executive Bob Finocchio and Les Denend, refined what was labeled the Renaissance Plan for the company in 1990. This trio, which would later be dubbed "the three amigos," felt that the company should move away from a diverse product lineup that included computer systems in favor of a streamlined focus on products for data networking. Benhamou, with his study at the French Grandes Écoles in Paris, his immigration from Algeria to France and, ultimately, the U.S., seemed well suited as the Renaissance man behind the Renaissance Plan, as the *Mercury News* and others called him.[47] And the board ultimately settled the issue of succession by naming Eric CEO on September 28, 1990—after a few zigs and a couple of zags.

The New Echelon in Leadership

Ultimately, Bill Krause's exit was sparked by this evolution, or revolution, in 3Com's product direction, a desire to move away from a systems and computing ecosystem, rather than the internetworking vision that might mirror what competitors like Cisco were striving for. This was coupled with the fact that people were becoming frustrated with Bill's leadership style. If you lose the support of your team, you've really lost your ability to be effective. It was clear that Bill was losing his support over a combination of his abrasiveness as well as technology vision. After hearing from Debra Engel[48] about the concerns his

47. Wolf, Ron. "3Com: the next generation." San Jose Mercury News. October 1989.

48. The genesis of what amounted to a small palace coup involved a limousine ride from a 3Com At The Half offsite management event in late 1989 at the Sonoma Mission Inn, in which Eric, Bob Finocchio, and Les Denend (the three amigos), discussed their grievances and hatched plans to create change within the company. Debra was also in attendance. In a stroke of luck,

team was expressing, Bill sponsored the process to find his own replacement, a search that ultimately ended with Eric Benhamou becoming CEO.

At the same time, Bill also recalled his 10-year timeline plan, to retire and make way for the next generation of leaders and that technology disruptions can occur in a similar cycle. Bill noted, "Bob and I had realized the mistake of not recognizing that technology developments were disrupting the market for networks of PCs and a new market for networks of networks was emerging. It was this key factor that caused me to push for Eric Benhamou, with his roots from Bridge, along with the board of directors, to be my successor. This also represented a great case study in the 'innovator's dilemma,' and how Bob and I missed the transition from networks of PCs to networks of networks." Bill summarized, "I was focused on doing the absolute best job I could of leading the transition—it was always about the best interest of 3Com and all 3Comers."

3Com executives universally expressed their respect for, and even amazement at, Bill's leadership during that pivotal decade: "Bill was very strong-willed, but great about training leaders, and building a strong team;" "Bill was great in front of an audience." And, as one person told me, "All of them were hugely dedicated to the company and it was their family." Another noted how Bill's dedication to 3Com was extensive, "He would spend the weekend thinking about everything at the company. He was brilliant at thinking about things that weren't working, and no matter how well a company is doing, there is always something that's not working."

Eric said, "Bill deserves a lot of credit for having built a very strong professional culture in the company. The company that I inherited from him was a company that was well run and a company that had a very strong sense of integrity and a very good values system. These things continued through the entire history of the company."

To this day, 3Com executives gather for lunch several times a year with Bill Krause, Eric Benhamou, and others in the Bay Area in Northern California. They have remained true friends over the decades, a testament to the strong bonds of their relationships.[49]

Before hiring Eric, the board (which included Bob Metcalfe and Bill Krause) had pursued a lead to replace Bill with Ken Oshman, one of the founders and the "O" of ROLM, and the CEO of Echelon at the time. Jack Melchor had suggested Ken in an earlier board meeting. Jack served on both ROLM and 3Com boards. Bill knew Ken and approached him while vacationing in Hawaii

---

by taking the limo ride, the three amigos and Debra avoided some box lunches that caused digestive problems for many attendees taking the group bus back home from Sonoma. Over 100 executives attended the event.

49. Other alumni also told me about frequent reunions. For example, Melissa Weiksnar, a Synernetics co-founder, holds an annual 3Com/Synernetics/Star-Tek/Chipcom reunion annually.

over Christmas 1989. A deal had been reached and everything was in place to make the announcement that Ken Oshman would be stepping in as CEO. Debra Engel remembers how the drama unfolded:

> We had the entire communication plan in place with the press releases ready ... We were waiting in the conference room for Oshman to come in before he met all the employees. After waiting for a while, everyone in the room became nervous wondering where Oshman was, or if perhaps he got into a car accident on the way. He finally called Bill and said, "I'm not coming." It wasn't going to happen.

Recalling that moment, Debra was amazed at how word did not leak out. Debra and the three amigos requested to meet with Ken Oshman to understand why he pulled out of the deal. He told them, "I was actually on my way. I've always learned to trust my gut, and my gut said no." Since part of the deal to sign on Oshman involved 3Com buying Echelon, Debra wonders if the reason he couldn't follow through was because "he was selling out Echelon in the process." She felt that if he thought the deal would result in harm to Echelon, "he couldn't justify it."

And what if Ken had taken the CEO role? He had a deep understanding of telecommunications networks from his ROLM days, and later on, one of his close friends, Dick Moley, would go on to help run a company called StrataCom, a frame relay switch company later bought out by Cisco, which became one of the drivers of a fateful acquisition for 3Com. 3Com's outcome may well have been very different under Ken.

Ken Oshman passed away in August of 2011 and Silicon Valley lost one of its best and brightest. He was gentle, understated, and mentored everyone who worked for him. Ken would walk through ROLM headquarters offices leaving late at night, cigar in hand, unlit, with a twinkle in his eye and a kind word for those working late in the office. As Keith Raffel, who was assistant to Ken at ROLM, said, "Working for Ken was like going to the best business school you could imagine".[50]

Echelon's last CEO was Ron Sege, a veteran of both ROLM and 3Com. Echelon develops open-standard control networking platforms, and designs, installs, monitors, and controls industrial-strength "communities of devices." Echelon's technology platform is embedded in more than 100 million devices, 35 million homes, and 300,000 buildings worldwide. Echelon was purchased by Adesto Technologies on September 14, 2018, for $45 million, an anti-climactic conclusion to Echelon's 30+ years storied history.

---

50. Markoff, John. "Kenneth Oshman Who Brought Fun to Silicon Valley Dies at 71." NYTimes. August 2011.

Three Amigos and Another Horse Race

After Oshman stepped aside, the board needed to make its move on a new CEO. Bob Metcalfe threw his hat in the ring, along with Les Denend and Eric Benhamou. Sales executive Bob Finocchio never fully entered the race, feeling that Eric had the technology vision needed.

Eric shared the board's frustration with 3Com's current product strategy, particularly given the data networking perspective he'd developed at Bridge. Les Denend (a retired Air Force colonel, combat pilot, special assistant to Zbigniew Brzezinski in the Carter Administration, and McKinsey consultant) had been hired to create a federal government systems kind of business, an arena he knew well, having spent twenty years in the military.

Bob Metcalfe couldn't garner the support he needed. Howard said it succinctly, "The board of directors, I have to believe, are a competent group of people, and I'd also come to the conclusion that we were going to have a new CEO. It was very clear that that group of people were not going to consider Bob Metcalfe or me for that job. So, Bill Krause, as the clock turned full, was going to be replaced and Howard and Bob were not going to be in it."

Bill Krause felt that "Eric had the network of networks background plus an insatiable drive to learn what he needed to learn."

To entertain alternate outcomes in a parallel galaxy—what if Krause had been less abrasive with others on his team? What if the three amigos had been Charney, Metcalfe, and Benhamou? We will never know, but we can surmise that Charney's departure and his resultant next startup (with assistance from Metcalfe) contributed to 3Com's ultimate demise. More on that later.

Les and Eric were first interviewed separately, and then later, side-by-side in front of the board. Board members took turns asking each of them questions in a sort of death round. "First Les would have to answer and Eric would respond. Then Eric would have to answer," according to Debra Engel. In the end, they wanted to keep them both. Weighing in on the outcome, Finocchio also noted, "I did not have the experience to be 3Com's CEO. Eric had the vision. Les had all the leadership presence." Les said, "Looking back, I'm surprised I did not withdraw, Eric was the guy. Thank the gods that the board had the wisdom to appoint Eric." Eric won out. He was named the new CEO of 3Com in the fall of 1990.

Eric pointed out that the board was not just selecting him—they were also voting on the product strategy going forward. There hadn't been a clear understanding that local area networks and wide area networks (LANs and WANs) would fuse together to create a massive and important network fabric, one that would span the entire world. Within a few years, companies like Netscape would spark consumer and business interest in the World Wide

Web, the demands on the underlying Internet, and an endless parade of new ideas that changed virtually every aspect of our personal and professional lives.

Eric recalls, "We had a close working relationship with Netscape Co-Founder Marc Andreessen, as Netscape CEO Jim Barksdale was on our board, and he invited me to the board of Netscape. We talked together about where the networking industry would go, us focusing on infrastructure, they were focusing on the browsing platform, but the two had to be meshed together."

Teenage Angst

In 1990, 42 percent of Americans had used a computer. Impressive for a product for which, less than four decades earlier, IBM's President Thomas Watson once famously predicted would have a global market "for maybe five computers." (Some insist that remark has been taken out of context; regardless, there's an underlying truth to it.) But that impressive technological adoption was about to be overshadowed by an even more impressive one. The 1990s would bring a transformation of the networking world that would also lead to a transformation of virtually every aspect of the global economy and daily life in the developed world.

While working on networking computers at CERN[51] in the late 1980s, British scientist Tim Berners-Lee began to discuss the idea of a hyperlinked information system with his colleagues. The first implementation of these ideas emerged in 1990, with the creation of the World Wide Web and the first web browser. Several other communications systems had recently been launched commercially—Lotus Notes launched in 1989, AOL Mail informed people, "You've got mail!", and kids used AOL Instant Messenger (AIM) to pass notes that same year. With these, the first Internet service providers emerged.[52]

With the launch of commercial services on the World Wide Web, and dot-com companies offering everything from products (like books, Amazon's sole initial product line) to services (remember travel agents?), and new, more ethereal, offerings like social networking, the world would be radically altered. This market evolution created a massive opportunity for companies to make the products to fuel this online revolution, which relied on the underlying Internet, a massive network of networks that would connect millions of

---

51. CERN is a European research organization that operates the largest particle physics laboratory in the world. They house a laboratory of 2,500 scientific, technical, and administrative staff members. In addition to particle accelerator research, they have helped with computer science including Internet technologies and the World Wide Web.
52. Pew Research Center. "World Wide Web Timeline". March 2014.

computers together globally. Those who made the right bets won out big; those who bet wrong were left only with massive headaches.

Chapter 9: The Mother of All Networks

Riding the World Wide Web Wave

As the World Wide Web and Internet boom began to build up steam, it sent shockwaves across almost every sector of the economy. Certainly, it helped to propel personal computer sales; between 1990 and 1997, PC ownership in the U.S. rose from 15 to 35 percent. While only a few million people used online services in 1990, that number changed dramatically in 1994 with the release of mainstream web browsers like Mosaic and Netscape Navigator.

New online services, like email, became overnight successes. That drove a wave of corporate development and acquisitions, including Microsoft's purchase of Hotmail, with the clout of online-native companies like Amazon, eBay, AOL, and Yahoo! growing rapidly. Related markets for other kinds of information technology were also changing and growing and changing again—at least for those companies able to keep pace with the shape-shifting nature of the information economy.

More broadly, capitalism itself seemed to be on a roll. Overall, the economy of the 1990s demonstrated consistent prosperity, due in no small part to the advent of the Internet and the explosion of technology industries that came with it. Personal incomes doubled from the recession of 1990, and personal productivity was up.

And of course, the booming demand for online everything coupled with a booming economy spelled out massive opportunities for the companies that provided products to support the fast-growing online infrastructure—one in which every component needed to be upgraded almost as soon as it was installed. Consumer and corporate modems—the devices that sounded a familiar chorus of annoying beeps as you connected to an online service over analog phone lines—were getting faster and faster. This put pressure on the online services to install equipment that could support those faster transmission speeds.

One of the market leaders in modem products (and related technologies) was US Robotics (USR). And while 3Com continued to have strong sales and some major hit products, it was still missing out on key markets. Within a few years, the company would target USR as its next major acquisition.

Meanwhile, in the years leading up to that acquisition, 3Com's financial numbers were strong. By the second half of 1990, 3Com had refocused its efforts and a turnaround of sorts was underway. The second half of FY 1990 saw 3Com with increased sales, up 15 percent from the first half to $224 million, with earnings per share doubling in the same time period. Internetworking was reasserted and the NetBuilder line, a single protocol router, had been announced. On the downside, 3Com had missed the market window for multiprotocol routers (devices that would work with competing network protocols), conceding it to Cisco.[53] As Matthew Kapp, a key sales executive in the Asian and European regions recalled, "Cisco was now literally at the epicenter of the new network effect of 'routers breed routers', which when well executed with the CIO and enterprise accounts, would lead to a familiar path of winner takes all."

However, the Ethernet board business remained a key focus for 3Com. While the analyst community thought of these as low-margin commodity products, the adapter business was quietly contributing heathy margins that enabled the investments needed to develop its more sophisticated products. This was downplayed by the company for many years—if no one asked, then no one volunteered to share how successful the adapter business continued to be for 3Com, as the company didn't wish to divulge the lower margins that the Systems products had, although analysts eventually figured things out for themselves. The company's annual reports and 10Q filings did not disclose this level of detail, and it was never volunteered.

With the "cash cow" adapter business generating hefty and consistent profits, Bob Finocchio and Eric could focus on filling in some missing pieces needed to serve the market for enterprise systems. In 1990, Bob Finocchio moved from VP of Sales, Marketing, and Service to EVP of Field Operations, Les Denend was overseeing product operations before he left the company shortly to go to Vitalink, and Alan Kessler took over the Distributed Systems Division.

As part of Eric's Renaissance Plan, 3Com announced the 3+Open Connection to NetWare in 1990, a clever and graceful way to pivot a messy exit from the 3+Open partnership with Microsoft into a new (and competing) market opportunity. The company also enhanced the NetBuilder line of internetworking products and added remote bridging and local and remote routers (more on NetBuilder and remote routers later on). The adapter business was going strong, having finally launched the emerging 10BASE-T twisted-pair compatible adapters for telephone wire environments, a key development in the marketplace that had gotten bogged down with the IEEE standards group for over a year.

53. Breidenbach, Susan. "Top 3Com execs talk strategy at briefing". Network World. August 1990.

From Science Fiction to Tech Fact

To help the company find the elusive missing links in its product mix, another colorful and brightly gifted thinker joined the team to contribute in a variety of leadership roles within Bob Finocchio's Network Systems organization: Jeff Thermond. Jeff became interested in technology while reading science fiction. His boarding school had access to Dartmouth's data center, connected via a General Electric (GE) network. Coincidentally, Jeff ended up in a GE management training program, eventually working for Tymshare in Chicago, which was bought by McDonnell Douglas. Jeff eventually accepted a job in Southern California with Bridge, where he led a successful regional sales team.

Recognizing Jeff's other talents, Finocchio recommended that he (along with Roy Johnson and other business savvy folks) be moved into product management and marketing roles.[54] Finocchio was frustrated by misdirected, ineffective marketing, that didn't meet the customers' needs. He felt that Jeff and the others would help make product decisions more effectively in their product management or marketing roles.

In 1990, Jeff Thermond stepped into Alan Kessler's shoes as Director of Marketing, just as the company had not only decided to buy their way out of the Microsoft relationship, but also to terminate the 3Server and 3Station businesses. This was a fortuitous time to exit from those businesses, as looming competition, such as Compaq's ProLiant line of servers, were cheaper and better.

However, while it may have been the right decision for 3Com, it was going to cause grief for some 3Com customers. As Thermond explained:

> The executives had this realization saying, "Hmm, if we pull the plug on the product, and it doesn't work out, the hard-core customers loyal to 3Com that bought into the server strategy both left and right—that might get screwed by this deal—will darken our reputation forever."
>
> So at the end of the day they decided, "No, what we're going to do is gradually exit the business and give customers a path to get off 3Server gracefully." At this moment in history, I had a young product manager that suddenly looked at me and said, "You know Jeff, I know I wasn't supposed to do this, but I grabbed a couple of engineers and we ported NetWare on to the 3Server." We called Novell and

---

54. Roy Johnson opened 3Com's first sales office in Australia, managed the Asia business from Hong Kong, and later became VP and GM of 3Com's Home Networking and Network Management.

> said, "Okay, we want to bury the hatchet and do a transition plan for people who have 3Com hardware and servers, so they can get special arrangements with you guys and special deals to adopt NetWare," and they said, "Sure, we'll do that all day long!" So, we did that, plus there was a bonus program if these engineering milestones were met to properly port it over to NetWare, which we did hit.

Sibling Rivalry in the Community

Starting in the 1980s, five companies emerged as the Big Five in networking—3Com, SynOptics, Wellfleet, Cabletron, and Cisco. Thanks to the common Ethernet standard, these and other Ethernet suppliers collaborated, sharing knowledge, OEMing products from each other, and filling in complementary products to their product line, and customers benefited from this as well—they could get buy 3Com adapters, SynOptics hubs, and communication servers from Bridge Communications. That said, these companies were extremely competitive, and differentiation became increasingly important.

SynOptics, an early networking hub leader, emerged out of Xerox PARC. Co-Founders Andrew Ludwick and Ronald Schmidt met in 1983. They saw the potential for IBM's approach to networking devices together: using the existing, ubiquitous telephone cables to connect devices, with what they called a Token Ring design. The Synoptics hub wired the Token Ring as a hub topology, whereas Ethernet was running at the time on a linear bus topology, and required the installation of coaxial cables to connect all the devices in a network. Xerox provided financial backing for Ludwick and Schmidt, and in June 1985, their new company SynOptics began work on LattisNet, a concentrator/hub designed to support Ethernet in a twisted-pair configuration, which launched in 1987. They brought in $6 million in revenues in 1987, which exploded to $40 million the next year, when they also went public, adding another $20 million to their coffers. 3Com OEMed SynOptics gear and used it in their own campus, up until its purchase of BICC Data Networks.

The company also later introduced their Network Management System that controlled intelligent hubs and other physical LAN components, which helped differentiate their product line. Like 3Com, SynOptics wanted to avoid being pigeonholed as a lowly provider of commodity products. (Network designers think of the various products in that universe as being part of a TCP/IP protocol stack in which the first (lowest) layer of the network includes devices like adapters and non-intelligent hubs, with increasingly sophisticated products filling the higher layers of the stack. Those higher layers are often where the highest profit margins can be made, while the lower levels—which may have been innovative and commanded high margins initially—quickly become low-margin commodity products.) SynOptics revenues continued to spike, to $77 million in 1989 and $176 million in 1990.[55]

SynOptics got into Ethernet switching by OEMing a switch from Kalpana, a company later purchased by Cisco. To graduate from only playing at the lower levels of the technology stack, and after Cisco discontinued their OEM arrangement selling to SynOptics, SynOptics countered by merging with a routing company called Wellfleet Communications in 1994, in a deal valued at $2.7 billion, and together formed Bay Networks. In 1998, the merged company was in turn sold to Nortel in Canada for $9.1 billion, forming Nortel Networks.

Interestingly, as it was described by others, when Ron Schmidt was still at Xerox PARC (before co-founding and serving as CTO for SynOptics), he had offered what would become SynOptics's hub product to Bob Metcalfe. At the time, Metcalfe's reaction was that Ethernet was never going to get so big that a product like that would be viable. Too many alluring opportunities at the time, perhaps?

A showdown among the network powerhouses was looming, and there was one more major player that would be part of the quarterfinal bracket. Around the same time SynOptics was getting off the ground in 1983, a company called Cabletron was founded in a Massachusetts garage. The company was founded by Robert Levine and Craig Benson, who later became New Hampshire's governor. The company's initial success was in providing products for the 10BASE5 Ethernet market, later producing hub products as 10BASE-T became more prevalent. They later partnered with Cisco to co-develop a layer 3 router. The company would take in over $1 billion in revenue in 1996, but as sales began to wane, Cabletron was reorganized as a holding company for four separate networking companies, none of which remain independent companies:

- Enterasys Networks, which ultimately became a part of Siemens Enterprise Communications, was later acquired by Extreme Networks in 2013.
- Riverstone Networks, based on the asset of YAGO Systems, was a company they bought to move into the switched Ethernet business, and was later liquidated.
- Aprisma Management Technologies, later sold to Concord Communications and then acquired by Computer Associates (now called CA, Inc.).
- Global Network Technology Services, a network installation and management company, which dissolved in the dot-com collapse in 2001.

55. "History of SynOptics Communications, Inc." http://www.fundinguniverse.com/company-histories/synoptics-communications-inc-history/

One could say that Cisco had two businesses: selling product and buying talent. John Chambers adopted the practice of acquiring whatever expertise he thought they needed and leaving the details for later. However, Cisco didn't rush into acquisitions just for the sake of acquiring staff, products, or customers. In Ed Paulson's book *Inside Cisco*, Chambers articulates five things that directly impacted their decision to not merge with either SynOptics or Cabletron:

- Statistics show that 50 percent of large-scale acquisitions fail.
- Merging two rapidly growing companies slows the growth and momentum of both as they work out the inevitable details.
- Cabletron was perceived as being tech-driven where Cisco was strongly customer focused.
- Cisco estimated that 60 percent of the existing channel partners would move to other vendors.
- Future alliances would be more complicated.

History suggests that John made the right call. He followed a path similar to 3Com's acquisition strategy (notwithstanding its challenging mergers with Bridge and USR) of folding in smaller companies, like switch technology leader Crescendo in 1993 (which turned out to be a huge success), and Kalpana in 1994[56] along with Howard Charney's company, Grand Junction Networks, acquired in 1995.

Setting the Stage for Explosive Growth

In 1993, CERN announced that the World Wide Web would be free for anyone to use, while the National Science Foundation (NSF) allowed the elements of the Internet housed in academic institutions to connect with commercial networks.[57]

With the Internet's doors now wide open for business, the entrepreneurial world went just a little bit crazy. Venture capital had been in a slump, with few startups attractive enough to invest in. From an investor's viewpoint, the '90s were indeed a bubble—stocks were rising a bit too much and everyone took brash product and market risks. Whereas investments in the '80s were more based on being cautious with known markets, exhibiting less ambition, and yielding decent returns, the '90s were all about chasing huge risks for big returns. Many paid off, but some did not.[58] For customers and end-users, the payoff was unambiguous. Thanks to companies like Google, Cisco, 3Com, and all the others that made the Internet-connected

---

56. From Ed Paulson's book, *Inside Cisco*, on pages 137 and 291. As an aside, Larry Blair, one of the founders at Kalpana noted also that they were in negotiations to be purchased by IBM when Cisco found them.
57. Giampietro, Marina. "Twenty years of a free, open web". CERN. April 2013.
58. Neumann, Jerry. "Heat Death: Venture Capital in the 1980s". January 2015.

world possible, a huge percentage of the world's population was now mere clicks away from practically all the information in the world, and from each other.[59]

Grand Junction, What's Your Function?

Howard Charney gave Bill his notice in 1989 that it was time to move on. But his entrepreneurial bug remained. An article in *The Economist*[60] titled "Out of the Ether" described Howard and some friends batting around the idea of how to make Ethernet ten times faster, at a company to be called Grand Junction Networks. Though it was transmitting data at 10 megabits per second, the rapid evolution of personal computing meant that speed was fast becoming a bottleneck. Larry Birenbaum who served as 3Com's VP of Engineering in the early 1980s, shared his 3Com and then later Grand Junction experiences.

In the fall of 1991, the new company—Grand Junction Networks—began brainstorming on technology ideas. Howard, Bernard Daines, along with David Boggs, former 3Com employees Ron Crane, and Larry Birenbaum, met at Howard's home. Andy Verhalen was also part of the initial team. After looking at LAN technology, FDDI, and ATM, Larry Birenbaum posed the critical question, "Why don't we just make Ethernet go faster?" With inputs from Ron and Bob Metcalfe (an early investor), and armed with an existing FDDI design already done by Mario Mazzola, Luca Cafiero, and others at Crescendo, Larry found a way to drive Ethernet to 100 Mbps.[61]

The Fast Ethernet standard ultimately allowed for both Larry's approach and 3Com's to co-exist, but critically, 3Com and the marketplace failed to anticipate that Grand Junction was also developing a switch, and not just faster adapter cards. Many of Grand Junction's employees had come from 3Com's Adapter Division too, which may have added confusion, i.e., folks thought they were going to make a Fast Ethernet adapter. Grand Junction masked the details of their products in their fall 1992 press release and 3Com and others were caught unawares on the switching capabilities.

In 1992, Grand Junction officially kicked off with successful VC funding and sold $4.5 million of their gear. By 1995, they were on the path to an IPO when Cisco took notice, and wanted to fill out their low-end Fast Ethernet switching products. While Goldman Sachs was preparing the IPO, Cisco kept calling, determined to acquire the company. Eventually, Howard Charney and

---

59. Goel, Tarun. "1990s Technology Timeline: Massive Growth During the Decade of the 90s". BrightHub.

60. "Case History: Out of the Ether". The Economist. September 2003.

61. Larry Birenbaum noted "we proposed, evangelized and wrote the 100Base-X/XT/XF standard and chaired the associated subcommittee. Mario, Luca, and Tazio were not directly involved in FE, but were nevertheless pivotal to its creation. Their creation of the MLT-3 Physical Layer for CDDI (FDDI/copper), and the associated silicon, was directly appropriated for FE!"

John Chambers agreed on a $345 million price tag for Grand Junction (leaving behind another suitor, IBM). In two years, Howard had gone from being the CEO of another Silicon Valley startup to helping to run one of the most valuable companies on NASDAQ.[62]

In recounting the standards war at the time, Larry drew out the party lines, "HP, IBM, and AT&T were all opposed to us, but Grand Junction, 3Com, Sun Microsystems, and Bay Networks were allies in the fight. We were all promoting this Ethernet-based standard, and we were very effective with the organizations that drove the technical standards—even later on, when we were at Cisco."

Ironically, the Grand Junction team that had come from the top talent of 3Com (Howard Charney, Larry Birenbaum, Andy Verhalen, Jack Moses, J.R. Rivers, Richard Hausman, John Celii, and David Schwartz, to name a few) would land one of the important knockout punches to 3Com. But, let's not give away the story quite yet. The speed of Cisco adopting the Fast Ethernet standard to the chassis-based switches from Crescendo, gave Cisco the launch of the Catalyst family Ethernet switching products, which is covered later. (When Larry left Cisco in 2004, Grand Junction's products were generating $3.5 billion a year, and commanding 65-70 percent of market share for desktop switches. Despite having a small staff of about 700 people, Cisco's phenomenal salesforce and support was able to make the product line extremely successful.)

Chapter 10: The New Guard

From left to right:
Chris Paisley
Andy Verhalen
Bob Finocchio
Eric Benhamou
Debra Engel
Alan Kessler
John Hart

Back at 3Com, the new arrivals at the company were anxious to try a different approach. Eric's background in data networking at Bridge left him frustrated with 3Com's current product strategy, and ready to create something new. The company had Les Denend, with his military and consulting background, launch an operation focused on federal government sales. But as the examples of SynOptics, Grand Junction, and others suggest, there were entirely new product arenas to explore and exploit. In many ways, the choice of the next CEO had been

62. "Grand Junction Networks". Fast Company. January 1998.

a referendum on the status quo—stay the course with the original team, or shake things up with someone, and something, new.

"I think you need a different style of leadership at different stages. When I was given the opportunity to lead 3Com, it was actually the phase in 3Com's evolution best matched to my skill set. Bob Metcalfe was an outstanding evangelist-type of a founder. I was very driven to build on this foundation and direct the company towards a long-term goal through a consistent strategy. And I wanted to pursue a goal that was meaningful not just from a technology development perspective but also from the perspective of making a general contribution to society. It's a bit lofty, but I felt we were building an important modern infrastructure for the world by connecting all these computers and all the people behind these computers."

- Eric Benhamou, 3Com CEO

That something new—whatever it would turn out to be—would be produced by a new senior team. While the three amigos continued at the company after the "not really contested" CEO horse race, some members of the original team decided to call it a day including Bob Metcalfe, selling his stake shortly after leaving. He moved to Maine, bought a 150-acre farm and some other property, along with a lobster boat he named Enthusiasm. Metcalfe also took a year out to accept a fellowship at the University of Cambridge. He then became a columnist for InfoWorld, joined Polaris Venture Capital as a venture capitalist, and later began teaching at the University of Texas at Austin. (A favorite Bob Metcalfe maxim, although he denies he said it, was one that was etched in stone on the Kifer Road campus driveway: "The only difference between being stubborn and being a visionary is whether you are right.")

Installing Eric in the CEO spot helped set 3Com on a more competitive course in the 1990s, largely due to his vision of global data networking. While Eric's consensus style of leadership—and his expectation that everyone should and would act like grownups—did not always result in getting things done efficiently, he had nonetheless inherited a very strong team. As Debra Engel explained, "Eric was thrust into leadership. Because he had all these operational leaders under him who are very good at doing what they do, we all just pulled together and went into gear to carry out his vision."

Eddie Reynolds added how impressed he was by Eric's vision, "For the next two years, Eric set the clear vision of what the technology was going to be, the customers, the competitors, and we communicated this well to the media, the company and the 'survivor' employees." Eddie also helped put together a 15-point list of suggestions for Eric, ranging from public speaking skills improvement, to information sharing, meetings, and so on, and Eric was a great learner, picking up the skills brilliantly. Eric also went through great lengths to build bonds with his team, hosting ExecCom offsites for cross- country skiing, rock climbing, or just floating in the Dead Sea in Israel.

When asked what he felt were the key elements of 3Com at the time he was taking over, Eric said, "It started off with the vision and three letter acronym of global data networking, to create that fabric and needing all the pieces of it and describing what the world was going to be like once we're fully connected. The foundation was this strategic vision. In 1992, we had 90 percent of the same employees we had in 1989, but things were clicking very differently, we had articulated the right strategy for who we were, what the market was, and communicated it well, both inside and outside."

A New Bob in Town

Eric's objectives, and his leadership style, clearly took 3Com in some new directions. To help implement those changes, he had a team of relative newcomers managing different aspects of the company. One of the most influential of these newcomers was Bob Finocchio.

Finocchio arrived at 3Com after having led successful teams in ROLM's sales, marketing, and service organizations. He was one of the most influential executives at 3Com, and led in sales and general executive roles during the mid-1980s until the USR acquisition in 1997.

After attending Harvard Business School, he felt driven to lead a company someday—as they said at HBS, "Look to your left, look to your right, one out of three of you will be CEO someday." Finocchio's father was an old-fashioned, retail kind of banker with Bank of America, who Finocchio described as "operationally savvy, tough and gruff, and intimidating"—something that could also be said about Bob!

And, in fact, Finocchio started his career at Bank of America in 1976. One of the first bank transactions he observed made an impression on his future path in the startup realm:

> This older guy and two young people walked in one day, and he was there to guarantee a loan. The two young people looked very scruffy, and it was [Apple Co-Founders] Jobs and Wozniak. And the person who came in with them was [Apple angel investor and Co-Founder] Mike Markkula. And he guaranteed a $250,000 loan,

> which was probably the highest ROI transaction in the history of the world. It was a very exciting time in the Valley.

Finocchio noted, "I inherited a banking relationship with ROLM and I got to know the founders and Joe Graziano, their treasurer." Finocchio shortly left and went to work for ROLM and Graziano, in a credit management role for the quickly expanding ROLM distributor channel. Ken Oshman was impressed by Finocchio's work, and began giving Finocchio assignments, like, "Go buy this company." If Finocchio responded by saying, "I don't know how to buy a company," Ken would just reply, "Doesn't matter, take a lawyer, go buy it." The management team continued to test him out in different roles. By the time IBM took over ROLM in 1986, Bob Finocchio had taken on all sorts of sales roles and was running worldwide customer support. By this time, he decided he was ready for something new.

He considered helping Ken Oshman run the newly formed Echelon, and was also contacted by Informix's CEO Roger Sippel, but ultimately Bill Krause lured him into 3Com. Finocchio was initially hired as VP of Sales, Marketing, and Service in 1988, with Jerry Dusa running U.S. Sales and Cate Muether managing marketing.

Finocchio and his team had a complex sales proposition on their hands, with servers, software, file sharing, and more. That complexity was compounded by internal politics—3Com had just purchased Bridge Communications, whose team had wanted to create router products as a solution for connecting servers, but that approach was nixed. 3+Open was a troubled product, and the economics of selling servers and 3Stations, which were basically dumb terminals, was changing as PCs took over the market.

It wasn't that Bill Krause earlier didn't want to build routers—but he wanted to create them as software products. Router functionality was something you could program a 3Server to do, as a software module in 3+Open. Unfortunately, the solution they built was horribly slow, and didn't support many networking protocols. And, before it became clear that there would be one dominant networking protocol (TCP/IP), customers were increasingly demanding that their investment in routers work with whatever protocol might lie in their future: Novell's, IBM's various protocols, 3+Open, or something else that might emerge later on. The genius of Cisco's early product and marketing strategy was to offer multiprotocol routers, which supported more protocols than anybody else.

Nevertheless, in Finocchio's early years, the company experienced great growth in sales of $250 million to $385 million from 1988-1989. In 1990, sales growth continued at a slower pace, rising to $419 million, but then dropping to $400 million in 1991. Against this backdrop—with not just the decline in sales, but also the end of the ill-fated relationship with Microsoft—the company embarked on what would become known as its "crossing the desert" phase.

Finocchio reminisced the sentiment at the time, "We don't know how we win, we don't know how we beat Novell, we don't know how we beat Microsoft. We don't know how we win in the computer industry. We don't know how we win by selling disk drives. Software might be a great business, but it might not be a great business for us." Finocchio added that it was hard to explain to Wall Street what we were doing around this time. As it turned out, the metaphorical desert 3Com had to cross wasn't just declining sales and broken partnerships—it was also a widespread sense that there weren't any visible destinations on the horizon.

Personnel and Personality

Earlier in 1989, the same year Judy Bruner hired me to head up a new internal audit function, Eileen Nelson was hired to support Howard Charney as a Human Resources partner with Debra Engel, the VP of HR. But before Eileen actually came on board, she was reassigned to Bob Metcalfe, who told her, "I know you were hired, but I have no use for HR, so the more you stay out of my way the better." Eileen Nelson was fortunate that Metcalfe respected IQ points, a sense of humor, and a focus on the business side of HR. He eventually accepted her in her new role, and they went on to be great co-workers.

Eileen later helped HR manager Sue Gellen when she became business partner to Bob Metcalfe when he took over the marketing organization and planned to fire a number of people in his initial strategy. Eileen told him, "I know your idea is to put shockwaves into the organization, and make clear to everyone your expectations are different than your predecessor, but this is not the way to do it." Metcalfe listened. He could be provocative, but not in a cruel way, even when he did want to shake things up.

Eileen Nelson also hired an HR manager in 1989 that would go on to take her own place down the road as SVP of HR in the story of 3Com: Gwen McDonald. Gwen, an African-American executive who recently retired from Network Appliance as Executive Vice President of HR, shared that her grandmother imparted her with a critical philosophy, "You are not responsible for the color of your skin, your big eyes, or your nose. But what I will hold you responsible is for the character underneath."

Eileen Nelson moved over to support Eric before he became CEO. In comparison to Bob Metcalfe, she found Eric to be quiet, introverted, stoic, and somewhat humorless, but nevertheless, she learned to adapt her style accordingly. On the Myers-Briggs assessment[63], Eric was an introverted, sensing,

---

63. The purpose of the Myers-Briggs Type Indicator® (MBTI®) personality inventory is to make the theory of psychological types described by C. G. Jung understandable and useful in people's lives. The essence of the theory is that much seemingly random variation in the behavior is actually quite orderly and consistent, arising from basic differences in the ways individuals prefer to use their perception and judgment.

thinking, judger (ISTJ), which was an unusual combination for a CEO, which are more commonly command and control personalities, such as Bill Krause's extrovert, intuitive, thinking judger (ENTJ).

Unlike the more charismatic leading Metcalfe or Krause, Eric was often viewed as not particularly charismatic or inspiring to the non-technical employees—similar to the way many people would characterize German Chancellor Angela Merkel. And like Merkel, he was unassuming and an intensely private character. Being a strong introvert, he was difficult to read, and not one to partake in verbal "brainstorming". But, once he arrived at a decision, Eric Benhamou gave clear direction, and full support to his close advisors and confidants. He kept his team intact for many years, garnering ardent loyalty—again paralleling the outcomes one might observe if you examine Merkel's government.[64]

Eric demonstrated his personal principles of integrity, patience, and intelligence. Like those who exhibit a contingency/situational theory of leadership,[65] Eric adapted his approach as circumstances required, but his decisions were informed by a supportive, participative, and generally consensus-building process. His careful listening, caring, and empathetic response was apparent when employees were laid off in 1990. His Renaissance Plan unfolded with much success (at least at the time), providing a solid example of Eric's effective servant and principle-centered leadership styles.

In Rich Karlgaard's book, *The Soft Edge*, the long-time publisher of Forbes magazine espouses the philosophy that great companies need more than just the hard-edge items that goal-oriented executives focus on—capital efficiency, logistics, supply chain, cost, and speed. The softer-edged assets—like trust, smarts, teams, taste, and a clear company story—are also essential.[66] Eric demonstrated this soft edge pretty well—his team had immense trust, they were all very smart, worked together for many years (at least up until the merger with USR, as you'll see later), and so forth. And Eric was especially effective at creating a compelling and long-lasting corporate narrative, with him providing his vision of global data networking as he assumed the role of CEO.

Eric's personality was almost the polar opposite of hard-edge. He was extremely patient and rarely, if ever, lost his temper. Eric was very cerebral; once he reached a conclusion, he was unshakable in his belief. His decisions were firmly based on clearly articulated principles, rooted in integrity. When

---

64. I wish to thank Jon P. Howell, Professor Emeritus, New Mexico State University, author of *Snapshots of Great Leadership*, for his additional insights about Angela Merkel's leadership traits. (Jon is also my brother-in-law.)

65. Howell, Jon P., "Snapshots of Great Leadership". Routledge. 2013. Pg. 9. Contingency and situational leadership theories assert the most effective style must fit the situation.

66. For a good view of Rich's triangle for long term company success, see *The Soft Edge: Where Great Companies Find Lasting Success*, by Rich Karlgaard, Jossey-Bass, 2014, page 16.

Eileen Nelson later became the SVP of Worldwide HR, it was easy for her to anticipate how Eric would weigh in on most issues, whether it involved organizational structure, executive development, or how to treat someone not making the mark. He operated with a strong moral compass; other 3Com leaders did as well, but their ideas about how to exercise their principles were rarely demonstrated so consistently.

First Steps into the Desert

So, in 1990, the board asked Eric—not yet 35 years old—to take the reins of a $400 million company that was experiencing a slowdown in its growth. Morale was weak, there were a host of strategic issues to fix, and folks were not pleased. After an all-hands management meeting, one of the managers took the "crossing the desert" confidential reorganization business plan booklet that had been outlined that evening, and faxed the whole thing to Computerworld, a popular tech publication at the time. It may have helped that manager vent, but it wasn't very helpful to the company.

Around this time, Vitalink executive John Hart was recruited by Eric. John's famous demo of remote bridging at DECWorld helped drive success for both Bridge and his own company Vitalink. At 3Com, he helped champion the vision that 3Com could pragmatically implement. His saw that companies such as SynOptics and Cabletron had successfully captured the lower-layer hub technology, while Cisco and Wellfleet were competitive on the higher-level routing products. John suggested a new strategy for 3Com along the lines of, "Why don't we do switches, and focus on the small and medium-sized businesses?" While there was a sense of "you have to try" to compete at both ends, much of the GDN was from small and medium-sized business based products.

Another critical piece was a battle over the network management standards called 802.1, which also importantly included the oversight for bridging, that pitted the Goliath IBM against the Davids of the world like 3Com. John Hart noted that after he joined 3Com as CTO in 1990, the work of engineers such as Mick Seaman and Floyd Backes, from DEC, helped drive through what is called the IEEE 802.1 standard, which was favorable to Ethernet winning in the higher levels of networking, such as bridging.

Eric had painted a vision in which information technology customers would be building bigger and better global data networks. The market needed high-performance devices, including dedicated and reliable routers and switches. The vision was lofty, and it would take a lot of painful sacrifices to fulfill that vision. As part of the desert crossing, the workstation and server divisions would be let go. The relationship with Microsoft would be cut; that would tighten the company's focus, but would also entail cutting a big check to Steve Ballmer at

Microsoft to get out of the contract. And it would take two years to get to the other side of the desert. Eric recalled a most interesting meeting with Ballmer:

> This was by far the toughest negotiation I ever had to lead not only because I was dealt a weak hand, but also because Steve Ballmer was a very intimidating, boisterous, outspoken guy. He would come down to 3Com to meet with us and we would discuss Novell during these discussions and of course they were an arch rival competitor to Microsoft. The thought of Novell would drive him into such a state of rage that he would take his shoe off and he would bang the table with his shoe the same way that Khrushchev, the Soviet leader went to bang the podium at the United Nations. Novell would drive him crazy. He was a tough guy to negotiate with, but ultimately not a crazy man. And it was essential for us to cut the LAN Manager and OS/2 cord and to move on, otherwise we would self-destruct. We still ended up paying a fair amount to Microsoft but not as much as our full contractual commitment, which would have left us completely broke.

Financially this made good sense—3Com was able to migrate 3Server customers to Novell's NetWare. And without any R&D expenses on servers, gross margins on the remaining sales, fell to the bottom line as operating margins. And the adapter business continued to steadily ramp up. But the financial pressure to stay ahead of the competition on all metrics was intense. Cisco, with about $100 million in annual sales, was only a third as big as 3Com, but 3Com's sales were stalling while Cisco was ratcheting up.

There were other challenges keeping 3Com busy in the desert sands. The company needed to shift from microprocessors to RISC processors in their router. (Microprocessor chips had more flexibility, but that flexibility came with a lot of overhead. RISC chips, on the other hand, were focused for devices that were dedicated to specific tasks.)

By eliminating product lines that distracted from the overall vision, cutting 15 percent of the work force, refocusing on the people and products staying with the company, and armed with the new generation of custom silicon, 3Com began to regenerate its momentum by the end of 1992. Around this time, 3Com was also the first to demonstrate stackable Ethernet switching technology (an important innovation that will be discussed later).

The board's bet on a young, ambitious, and idiosyncratic entrepreneurial engineer named Eric from Algeria was paying off. Jim Barksdale, a Bridge board member who carried over to 3Com's board, understood which technology requirements would be critical in the 1990s, and was instrumental in pushing

Eric as CEO in that decisive board meeting. Eric may have been inexperienced in many regards, but he learned on the job about building shareholder value.

The NetBuilder II Edge

A critical component in implementing the Renaissance Plan was developing innovative products internally. To that end, Eric stressed how important the 1992 NetBuilder II launch at Interop was, saying, "We knew that we had to play catch up on multiprotocol routers. There was no way to be a key player without a strong multiple protocol router. It's just that ours had to be different—and had to leapfrog Cisco. The first 3Com multiple protocol router that we introduced at the Interop show in 1992 was NetBuilder II. This was our approach to leapfrog Cisco. We were the first company to combine RISC technology and ASIC technology together to create breakthrough performance at lower cost."

Eric also noted that John Hart deserves a lot of credit for inventing boundary routing that Jeff Thermond explains later on. The new products made it easier to deploy a highly-distributed wide area network (WAN) around several central sites, with more peripheral sites that could get by with much lower-cost boundary routers.

NetBuilder II was announced in September 1991 and was shipped by early 1992. Having required a significant R&D investment, it was revolutionary. The devices (priced between $10,000 and $33,000) combined a reduced instruction set computer (RISC) processor and a custom-designed application-specific integrated circuit (ASIC), which supported high speeds for interfaces like FDDI. Thanks to the ASIC, the discrete components could be shrunk. The product would later add Token Ring modules and expand to include multiprotocol routing.

The NetBuilder II with boundary routing solutions was well liked in the Asian and European markets. While NetBuilder I products had poor margins, NetBuilder II had very high margins and helped get 3Com back to the number three position in market share, behind Cisco and Wellfleet. Alan Kessler also contributed by getting the first perfect score ever for a product evaluation in Network World magazine (which was a big deal in the industry) for an FDDI to Ethernet router. The NetBuilder II was well integrated, reasonably priced, and really, really fast at FDDI to Ethernet. He smartly figured out a metric called "packets per second per dollar" and it just slaughtered the Cisco products.

As Jeff Thermond said, "All of a sudden, everybody looked up suddenly and said, 'Wow! 3Com is back!' We won the network business for all eighty colleges in Oxford University for their big distributed FDDI ring, which each college ran on Ethernet. Cisco ended up at one point offering them the equivalent kit for nothing, but their initial bid had been so high that they completely pissed off the purchasing people at Oxford. Europe was really good for NSD. We got 45 percent of our business out of Europe, about 15 percent out of Asia, and 40 percent from the U.S.; we actually had better sales in Europe that we did the U.S." Jeff also noted that they built it up to $365 million in annual revenue, with more than half from four customers—Microsoft, Columbia Healthcare, Wells Fargo, and Deutsche Bank. Another great European customer was The Boots Company (Boots the Chemist), which had the largest point of sale network in the retail world at the time.

Breaking Boundaries

Jeff Thermond noted that when he arrived from Southern California to run Product Marketing in the Network Systems Division in late 1990, he heard Cate Muether, a former Bridge marketing manager turned Cisco employee, speaking at a marketing association meeting, where she was commenting on 3Com's ninth place in the market share for bridges and routers. This prompted him to take action.

Jeff, based on his training at GE, quickly pointed out to his division GM, Alan Kessler, the need for product road maps, and embarked on establishing what he felt was the right product direction:

> The business opportunity was basically terminal servers, system protocols, TCP/IP stacks for Windows PCs and Macs (before Microsoft made the stack native in Windows 95), OSI gateways, and IP routers. As near as I could tell the part of the business that was dramatically underfunded was IP routing. So, I said we're halfway through a new communication server platform that's in terminal servers, which are now declared "terminal,"—this is a line of business we are going to get out of because it's all going to go to PCs and LANs. We're betting on the wrong protocol. I want to stick to IP routers. I'm canceling the rest and we're taking them out of the product catalog. We're betting this entire division on IP networking, and that's where it's going to go.

As an aside, but related to the importance of product decisions, Jeff Thermond also explained how Bob Finocchio and 3Com's other executives shared the ideas of business thought leaders:

> I give Bob [Finocchio] immense credit for discovering this book by Steven Wheelwright and Kim Clark called Revolutionizing Product Development.[67] Wheelwright had been at Stanford and then the

---

67. Wheelwright, S. C., and K. B. Clark. "Revolutionizing Product Development: Quantum Leaps in Speed, Efficiency and Quality". New York: Free Press. 1992.

> Harvard Business School, and that book was hugely influential on everybody at the ExecCom level; Bob made it the Bible. At one point I remember Bob bringing Wheelwright out to 3Com, in 1994 or 1995.

Wheelwright's book covers the cycle from predevelopment investigation, to the systematic development project stage, on to product introduction.

Jeff Thermond explained that Jeff Krause (at the time, the Director of Internetworking Technology) was working with John Hart, the new CTO on a technology that John had come up with called boundary routing and felt it would be a great product to sell. Thermond had found out that the product was costing almost 50 percent more due to high factory overhead costs. By running the factory at capacity, product costs would fall significantly, allowing lower pricing.

Also at the time, Jeff Krause, Joe Furgerson, and John Hart developed a clever marketing advertisement that showed a very complex router on the left side, and on the right side was boundary routing. This illustrated the critical value proposition that boundary routing solved, reducing expensive links and connections, greatly simplifying the routing logic.

John Hart explained their thought process in creating the Boundary Router:

> Why do you need to go through a million lines of routing code? Why don't we just skip it? We just won't route. We'll put the router at the other end, but we'll treat this as one big connector. Just use the link and such as a connector. And that was the Boundary Router. I said, "File a patent on this!" as nobody had done this. And, to explain how simple this was [compared to Cisco's big ugly boxes] I would share at seminars that "here's the stack of manuals to install Cisco. And here's ours you can fit on the back of a business card."

Jeff Thermond slashed the price of the existing router to $3,900, and made the top hardware priority something called Spectre, the internal codename for the hardware platform to support boundary routing. It came in various software flavors, and was a 1U high router that did not boot from a floppy diskette. He said, "We're gonna be able to sell this new boundary router now for $2,500 and make a ton of money because it's designed right." Bridge employees provided a very talented software team to help with writing the boundary routing software. They slashed the price, and started a war with Cisco on a front they'd never experienced. John Hart introduced his friend Steve Russell from Vitalink to help run engineering in 3Com's router business.

Jeff Thermond explained the boundary routing concept:

> The codename came from a novel by Fritz Leiber called *A Spectre is Haunting Texas*. It's a science fiction story I read when I was a kid that had a big influence on me. It was a great metaphor: "This is the Spectre that is going to haunt Cisco." And it was damn good we did it when we did, since the next week Cisco launched their 2500 Series which was exactly like the product we had built for SuperStack [where the NetBuilder I fit in], and because Cisco hardware in the early days was ugly as sin, not elegant, and not particularly dependable.
>
> Joe Fergerson and I got hold of Nick Lippis to do a study that validated that the cost of managing remote complex routers in a branch office is actually more expensive than the cost of buying the damn thing or running a wide area link. So, we could go out and say boundary routing will cut the cost of your wide area networks by 50 percent. This helped launch boundary routing tremendously."

Thermond also noted that they uncoupled the software and hardware pricing, such that there were various flavors of software you could purchase with the hardware. This enabled a lower price point entry for customers. Jeff explained:

> Our whole strategy on boundary routing was to get people to look at us first, for anything at all for WAN [wide area network] purposes. Their old choice was to buy an old NetBuilder I. The minute Spectre came out at the fraction of the cost of the old NetBuilder I, boundary routing took off and at the low-end of the product line, we could go toe-to-toe with Cisco. For the first time, ever.

Chapter 11: The Acquisitive Years

While there may have been some benefits resulting from the Bridge acquisition, building out a robust product catalog wasn't one of them. And during the desert-crossing years, product and channel development were intentionally very narrowly focused. That left the company with a lot of holes in its offerings and sales strategy—which ultimately drove a flurry of acquisitions.

3Com was trying to make up for lost time between the Bridge purchase and Eric's ascendance by scooping up future networking technologies from

We're Number...Two?

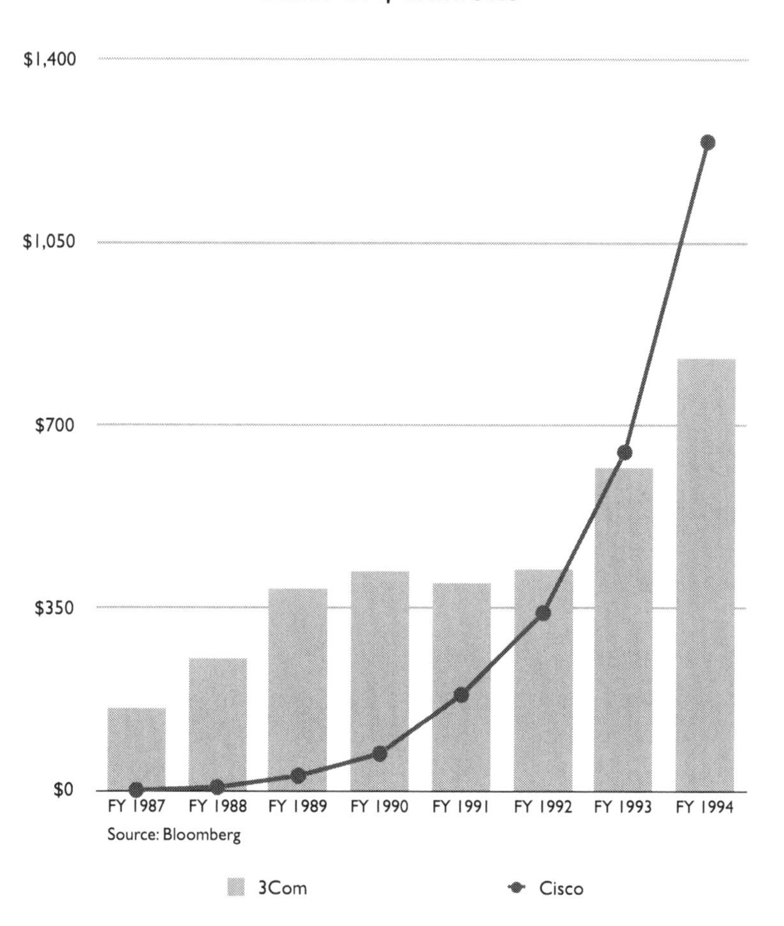

3Com had early sales success with its adapter-oriented strategy. Despite a variety of setbacks—exiting the computer systems market and a failed partnership with Microsoft—3Com embarked on a new emphasis on global data networking and outsold archnemesis Cisco until FY 1994. The acquisitions of BICC and Synernetics in 1992 and 1993 helped fuel the company's growth, but it was becoming increasingly clear that Cisco's focus on internetworking products within large enterprises would surpass 3Com's mass market products like adapters and stackable hubs.

smaller startups. The company's activities along with Cisco and others helped to fuel a broader networking mergers and acquisitions (M&A) craze.

In the 1960s, '70s, and '80s, there wasn't nearly as much M&A activity, and many of the deals that were done wound up being train wrecks. But in the booming '90s (a mild economic recession notwithstanding), M&A was suddenly on a roll, with tech companies like Lucent, Bay Networks, Nortel, and others taking part. M&A was no longer viewed solely as a tactic for adding products to fill niches or a mashup to scale up instantly, but increasingly also as a means to fill a company with skilled people who would bring talent to the bench.

This M&A activity may have also driven the growing pool of money for more startups. If an IPO wasn't in the cards, venture capitalists could always sell their startups to a 3Com or a Cisco. Later on, Cisco accelerated this strategy by purchasing hundreds of companies, and creating new models such as spin-ins, whereby they would invest in a few entrepreneurs up front and then later buy out the entity once products were successfully developed.

Slick BICC

Bob Finocchio, who was running sales, service, and marketing at the time, reflected on the time period that fueled 3Com's string of acquisitions, leading up to the fortuitous acquisition of BICC Networks. He remembered the conceptual pitch he and his teams would deliver to potential customers, "You know, someday every desk will be connected to a network ... every desk might have a PC on it." Most customers were skeptical at first, countering with, "No, why would you want to do that?" But ultimately, the powerful promise of this technology, especially in how it allowed the sharing of expensive resources, won customers over. Business was exploding.

In this context, 3Com was looking to expand its arsenal of networking products. NetBuilder was relatively underpowered relative to Cisco's offerings at the time. Finocchio recalled, "So, to some extent, we were at the right place at the right time. But, fundamentally, all we had was this really good adapter card business, and a somewhat struggling box business."

3Com was on the hunt for acquisitions. That search eventually led to their first offshore acquisition of BICC Data Networks in the UK for $25 million cash and $5 million in additional stock in 1992. BICC was a large conglomerate that built factories, buildings, and bridges, but they happened to have this small unit that made a hub. 3Com was itching to get into the multi-services hub (MSH) market to compete with the likes of SynOptics, Ungermann-Bass, and Wellfleet. Finocchio recalls the acquisition:

> We wanted to buy this company for its hub. We bought the company for almost nothing, because the sellers had no idea what they had and

> no idea what to do with it. I think it was about a $30 million-dollar business. They were located in Hemel Hempstead in England, and had a really good development team. Eric and I went over there, did the math, and fundamentally the cash we paid was no more than the revenue they were booking. So, it was almost free. And Janice Roberts came with that deal.

Jeff Thermond offered some additional under-the-hood details on the BICC deal. He recalled how the review of BICC's hub engineering revealed that they were vastly behind on their first release. The architecture was still being debated, and the actual product wasn't ready. At the same time, however, they had something—called Velcro—that they were almost embarrassed to bring out. Jeff was surprised that it looked a lot like the stackable hub that 3Com's Network Adapter Division (NAD) was about to launch, but it turned out to have a 40 percent lower manufacturing cost.

When Eric got in later on Sunday evening, he asked Jeff Thermond, "What do we have here?" Jeff explained that the news on the multi-services hub front was bad. "They are going to be far later than they imagine, and they don't even know how much trouble they are in." Eric was now convinced that the trip was a complete waste. Jeff then asked him, "You know that stackable modular hub that NAD is really excited about? What if you could have that with a 40 percent lower BOM?" Jeff recalls that Eric "leans forward and asks, 'What is that thing called?', 'It's called Velcro. By the way, it's uglier than sin, but we know how to bend sheet metal and these BICC guys have done a brilliant job—it's integrated a bunch of stuff that we are using discrete components for.'" At this point, Eric was intrigued.

Acknowledging that they had something good on their hands, Chris Paisley instructed the team, "Don't show enthusiasm on anything!", wanting Bob Finocchio, Eric, and Chris to keep their cards close. As the meetings with BICC were coming to an end, Janice Roberts, its general manager, came in looking uncertain. According to Thermond, Chris told her, "Great, thanks for showing all the stuff." Janice replied, "You're going to get back in touch?" Chris's response was another, "Thanks for showing us all the stuff." Ultimately, Chris did a great job negotiating the contract, including taking out some potentially expensive and dilutive stock option warrants that BICC didn't ultimately want (cash spoke louder than warrants). This was at a time when 3Com's stock may have been poised to soar, and was a good coup and savings for 3Com and Chris.

Janice became a critical addition to the 3Com team, an instance where 3Com was able to retain great talent. Janice was raised in the Cotswolds in England, hailing from the small village of Wiltshire, a very long way from Silicon Valley.

She ran Business Development, Marketing, and later, Palm until Robin Abrams was hired in 1999. She is known for being smart, articulate, quick, and ready to tell you how she sees it. Janice was an amazing deal-maker and a negotiator par excellence for 3Com, and a much-needed complement to Eric's wonkish engineering style. She was good at reading people and a natural at corporate marketing, while she may not have been as strong on product management—that's like saying, in football parlance, the wide receiver isn't as good at the offensive tackle position.

"Coming into 3Com, the executives were testing me and each other. In the UK, it's more a team sport; in the U.S., I found it to be an individual sport, much more competitive and fierce."

Janice Roberts, BICC SVP, 3Com SVP, Palm CEO

John Boyle, the marketing director for the Network Adapter Division, later worked for Janice in business development. They were a part of the Marketing Department, not Finance, which was unique at the time, and tailored them more towards identifying key markets or technologies, not just a financial target to meet sales impact projections. Boyle reflected that together, Eric, John, Janice, and Bob Finocchio nicely complemented each other's skills. Eric was the engineer, tech savvy, and strategic. Janice was tactical, people-oriented, and relationship-based. Boyle focused on positioning in the longer term, while Finocchio focused more on: "how does this help next quarter?"

After the BICC acquisition, Bob Finocchio asked Eric if he could run all the networking system products. Without much hesitation, Eric agreed. Finocchio wasn't an engineer, so this involved a big leap of faith. Finocchio assumed responsibility for the router division that Steve Russell was building, and oversaw BICC, which all combined became Network Systems Operations Group (NSOps). Eric told Finocchio, "You know, if you fail, I'll fire you." Bob said, "Yeah." Eric also said, "If this doesn't work, the board might want to fire me, but I'll fire you first."

BICC was the first step in an acquisition binge for 3Com. The company went on to buy Star-Tek for $50 million in 1993, a Token Ring company founded by Pete Williams, as well as Synernetics, a switch company based in Massachusetts. According to Finocchio, this allowed 3Com to have "a critical mass—or at least a hodgepodge—of some switching and a hub."

Finocchio recalls the closest he ever came to inventing a 3Com product. At a team dinner at his house in Saratoga, he explains how he and his team were inspired by his stackable stereo to develop the NetBuilder SuperStack series:

> I opened up my stereo cabinet. I had bought this component system from Sony. It wasn't racked, but they were stackable stereo components. I said, "Why can't we do something like this?... It doesn't matter what's on the inside. Let's at least make the form factor the same, the color the same, the naming architecture the same, and we'll work on the insides as we can. Maybe they'll be kludgey connections in the beginning, and maybe we come up with a better backplane later. But let's have stackable networking stuff."
>
> The BICC guys say, "Yeah, instead of making a card for a hub, we can turn it this way, and put a little power supply, and maybe sell it for 1,000 bucks instead of 10,000 bucks." And Pete Williams said, "Yeah, I can take my thing and do it that way." And we said, "We can repackage our low-end router, and we can do that too." And that was the beginning of the NetBuilder series [part of SuperStack]. No one else had stackables, no one else was going to attack the low-end of the business market.

After that dinner, the team began using that crude design to develop low-end stackable hubs, later also part of the SuperStack family. Following Ron Sege's advice (Ron had been moved to run BICC, later named Premises Distribution Division) to build a switch chip from a development team he found in Ireland, 3Com wound up with the first low-end, super cost-effective switch as part of the SuperStack line. The sophisticated hub allowed for all needed functions without the expense of a chassis, and without worrying about the power supply. Bob Finocchio recalls, "We owned the architecture, and we owned the silicon. And that really is what fueled the business. The Premises Distribution Division in the UK went from $60 million in revenues to several hundred million—and at the peak, almost a billion dollars. That was by far the most successful non-adapter business we had."

3Com gradually built a large portfolio of products. As Finocchio said, "They weren't tightly integrated, we had a lot of overlap, but it was clear we were ahead in low-end switching. We were never going to win in high-end routing." 3Com had also bet on Asynchronous Transfer Mode (ATM) technology, but it never took off.

Bob Finocchio was named EVP of Field Operations in 1990, and then EVP of Network Systems Operations Group (NSOps) in 1993, managing six product divisions/profit centers, including product development. In 1996, he was named President for 3Com Systems, which comprised 10 different divisions. There was

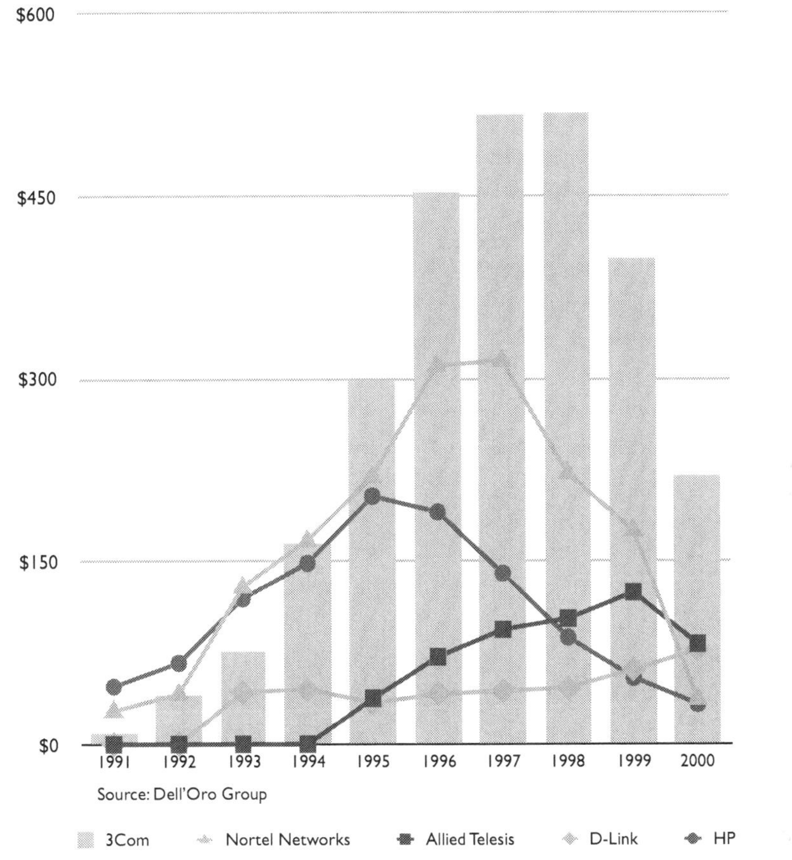

3Com dominated the stackable hub market in 10/100 Mbps speeds through the 1990s.

some irony in his earlier NSOps assignment. As 3Com's head of sales earlier in his career, Finocchio had been wildly critical of their product marketing and engineering, which won him zero friends upon his arrival as the business group leader. Yet Finocchio was smart enough to realize that he needed to change his style, and he quickly acknowledged to the group that his prior hyperbole was a sales and marketing tactic, eventually winning the team's support.

Finocchio worked hard, and expected his employees to do the same. He had a great sense of humor, but could also be extremely impatient, and was feared by some. On the one hand, he might shred a young analyst or manager that wasn't crisp and clear in their data or presentation. But he could also be an effective and instructive mentor. Nachman Shelef, who had helped sell NiceCom to 3Com, said that he had learned more about business, management, and life from Bob Finocchio than anyone else except his father.

Finocchio also recruited HP executive Ralph Godfrey in 1990, who later ran all of U.S. Sales, and stayed on until 2000. Ralph successfully implemented a two-tier distribution sales model. Distribution companies like Tech Data and Ingram Micro stocked 3Com's inventory and provided customer credit to integrators and re-sellers. This aided in the streamlining of mass market products such as adapters and the SuperStack line. Ralph recalled, "We kept the company in business—I had all the distributors—that was where all the money was coming in—while Finocchio, Eric, and Woody Akin (VP of U.S. Systems Sales) fought the systems battle."

Preceding this time frame, on the East Coast, the minicomputer industry was dying a horrible death—companies on the 128 beltway, like Prime, Data General, Wang, and DEC.[68] PCs and workstations were ravaging their businesses, and the systems engineers in that area ended up building systems networking companies. Synernetics was one such company.

A Noir et Blanc Decision

Around the time Jeff Thermond was made a VP, he was thinking about a less than optimal OEM deal that Tom Steding (VP/GM of the Network Systems Division) had made with Synernetics, (that Alan Kessler helped rewrite later). Synernetics had one of the first chassis-based LAN switches in the market, ahead of competitor Kalpana, which was later scooped up by Cisco.

3Com had a tradition that when somebody was promoted to VP, Eric and other members of the leadership team would have a congratulatory dinner at Le Mouton Noir, a fabled French restaurant in downtown Saratoga. A week before the dinner, Jeff Thermond had told Bob Finocchio what was on his mind at the moment: they needed to buy Synernetics right now.

Finocchio said, "You're going to drive Eric crazy. Are you planning on bringing this up at dinner?" Jeff said, "Yes, there is no time to wait." Finocchio warned him, "Then, you better really come in with your ducks in a row."

So, knowing the way Eric's mind worked, Jeff created a diagram, showing how things would play out if they bought Synernetics and—when you flip the chart over—what would happen if they didn't. Jeff seemed certain that Wellfleet would buy them within 90 days, and that the consequences of that would hurt 3Com. Jeff also pointed out to Eric that, "Every fucking person on the Microsoft campus was connected to a Synernetics switch, every one of them sold either through the OEM agreement or directly by Synernetics and that wasn't something you want to see Wellfleet[69] take over." This offered

---

68. Boston's decline in computing is covered well in the article, . "Tech's Lost Chapter: An Oral History of Boston's Rise and Fall, Part Two". https://bit.ly/2TCnaB2

69. Paul Severino, the successful CEO of Wellfleet, reiterated to Leon Woo how much of a mistake he made by essentially letting 3Com buy Synernetics. He felt the Wellfleet direct

customers great price/performance; folks that were buying routers for local area routing really just needed switches, so the Synernetics product was half the cost of the Cisco ten megabit Ethernet port router, but offered the same logical level of control.

It was going to be a gigantic business, and a big game changer. As Jeff Thermond tells the story:

> We're at the dinner and Finocchio says, "Eric, Jeff wants to talk to you about something." Eric had just given me my business cards with my new title on it and that was a very gratifying moment. He looked at me and said, "What's your idea, Jeff?"
>
> I said, "We need to buy Synernetics right now." Eric said, "This could be the shortest tenure of a vice president in company history." I told him "I understand that you're skeptical, I came prepared for that," and I showed him the diagram explaining what happens if we buy them. Eric looks at it, he's kind of doing this thing with his mustache, and says, "OK, you got anything else?" I say "Flip it over. That's what happens if we don't." He flips it over, he looks over at Finocchio and he says, "What code name are we going to give this project?"

The Synernetics Secret

Synernetics at the time was the number one Ethernet switch, ahead of Kalpana, which Cisco bought in the arms race of networking. Prior to 3Com's purchase of Synernetics, an interesting 3-way relationship developed between Microsoft and Synernetics and 3Com that also moved the needle toward acquisition. Based on some strategic work done with Boeing, resulting in critical introductions to Microsoft, Don Seferovich, Director of Business Development for Synernetics, sold Microsoft on Synernetics products for their worldwide backbone. "Our ability to focus with a laser beam, our engineering and tech support of their network, with plug and play solutions—was how we 'drove' our huge accounts like Rockwell, Boeing, and Microsoft."

But they were asked by Microsoft to OEM their product to 3Com with the idea that 3Com would be able to support the product better based on Synernetics' small size. But afterwards, 3Com's support faltered. Microsoft found a clever way to evade the 3Com OEM contract and buy directly from

---

salesforce, similar to Cisco's salesforce, would have been a better match for Synernetic's product line than 3Com's channel focused salesforce. For more info on Paul's early success in routing at Wellfleet in the early 1990s, see the New York Times 1993 profile on Paul: "After This Many Hits, It Can't Be Luck." https://nyti.ms/2T0w7iF

Synernetics by cleverly asking for new product requirements on Synernetics, an alteration that changed the OEM product with 3Com. This put 3Com and Synernetics at odds with each other prior to the acquisition. 3Com would thus recoup these lost sales by acquiring Synernetics.

Another benefit of the deal was that it helped anchor many 3Com accounts. Wherever the company sold a dollar's worth of router product, it could drag along two to three more dollars in Synernetics switches, branded as LinkBuilder. Wells Fargo's second project with 3Com was replacing their core products with those of Synernetics. And the same thing was happening at Columbia Healthcare, which was a $100 million-dollar account at the time. (Columbia Healthcare was a $9 billion-dollar company that put 3Com in charge of running their entire network.) Dave Tolwinski, a marketing executive from Synernetics, took over for its CEO, Allan Wallack, who exited upon the acquisition, making Tolwinski the general manager for the Synernetics division within Network Systems Operations Group.

Tolwinski shared his thoughts while he was snowed into his Massachusetts home during the "bomb cyclone" of January 2018. "3Com had a great brand name, and they were in a position to receive the benefit from the commoditization of networking products. The industry was really pretty broad, lots of networking products at that moment in time, and 3Com used the startup community as sort of an adjunct development facility."

To help manage the imminent period of hypergrowth, Jeff Thermond brought in a Bridge marketing guy named Clint Ramsay to help Tolwinski's division. According to Tolwinski, he didn't think he'd have that much trouble making $50 million the next year, since they'd done $30 million the year prior (with $18 million being the OEM deal with 3Com). But Thermond told him, "No, no, you're going to make $100 million next year." Eventually they got the business up to $500 million. Tolwinski and Thermond also worked together to solve the channel conflict between the 3Com and Synernetics salesforce. It was the perfect product for the 3Com salesforce. They were selling a LAN device which was a comfortable product for a more LAN- versus WAN-centric salesforce. 3Com had a huge price advantage and it took Cisco some time to come out with their Catalyst Switch line afterwards.

Less Chipper with Chipcom

3Com went on to acquire public company Chipcom, another hub company, in the summer of 1995 for $700 million in stock. At the time, Chipcom was the sixth largest networking equipment maker, and had strong ties to IBM. That quarter, Chipcom had sales of $71 million and annual revenues of $267 million. In contrast, 3Com's trailing 12 months' revenue was $1.3 billion. This was 3Com's biggest buyout to date. 3Com stock was trading in the $70's range at this time and they had a market capitalization of $5 billion.[70]

Many felt that 3Com was bamboozled. While Cabletron also wanted to purchase Chipcom, 3Com prevailed. In the end, 3Com's move may have been more of a competitive move to stop someone else from taking over the company. But it was yet another architecture to absorb, and IBM's demand for their product was shrinking at the time of the purchase. And hub technology, with its shared bandwidth was being obsoleted by switching technology with its dedicated bandwidth, as the switch technology costs were rapidly declining. Why would a customer buy a lower performance, rapidly obsoleting hub product when a higher performance switch product was available at a similar price point?

Leon Woo explained, "I did the technical due diligence with Menachem Abraham [Chipcom SVP of Product Development, and later Lucent CEO]. He was very honest and could not lie. But when I asked the hard questions, he did not answer, which was an answer! Their ASIC development was in trouble. They had acquired a small switching product from a DEC spinout called NexLAN which was a point product to cover their development gap. That product had a lot of overlap with existing 3Com products." And, culturally, the acquisition would be a challenge for others. Chipcom's CEO, Rob Held, was a former submarine lieutenant and its R&D leader, Menachem, was a command and control oriented Israeli. Unsurprisingly, their style was a departure from 3Com's federalist state model.

Bob Finocchio was initially supportive of the deal. But after more due diligence, he concluded that the backplane development did not exist and that their stated sales backlog was "bullshit." Their product plan for next-generation products was almost non-existent. Before the deal was finalized, Finocchio withdrew his support saying, "Look, this is not what we thought it was. Yes, we'll have to pay a break-up fee, but I think we're far better off walking away from the deal." Chris Paisley agreed with Finocchio.

In some folks' view, the deal was consummated by the force and influence of the investment bankers.[71] In their business, the real client is the deal itself—not necessarily the buyer or seller, contrary to what we might believe. Reluctantly, the board approved the deal.

Of the deal, Dave Tolwinski said, "The Chipcom product pipeline was dry, so we didn't get any product help. There also was a lot of product overlap. Regarding the salesforce, there was conflict in the sales territories, and it

---

70. Bloomberg News. "3Com Agrees to Purchase Chipcom for 700 Million". New York Times. July 1995.
71. The investment banker for this transaction was Frank Quattrone, one of the top investment bankers in Silicon Valley, and referred to in the book called "The Prince of Silicon Valley: Frank Quattrone and the Dot-Com Bubble", written by Randall Smith. Quattrone and his team helped 3Com as well as earlier company Bridge Communications extensively.

became a complex mess figuring out how to compensate them and who was responsible for what areas."

Chipcom was based in Southborough, so the company thought it could align itself with nearby Synernetics in Boxborough, but the technologies and channels turned out to be very different from what was expected. And when the 3Com East Coast team finally came out with a competitive product called CoreBuilder, it was one to two years late and light on features. The first CoreBuilder 3500, was their first low cost router, and competed with Cisco's lower end Catalyst line for a short time. CoreBuilder 9000 featured a backbone layer 3 switch that could handle ATM, Fast Ethernet and Gigabit Ethernet traffic, and would attempt to compete with Cisco's Catalyst Fast Ethernet switching product. We'll revisit a fateful decision that Eric made on the CoreBuilder product line later on.

During the 1990s era, the core network products related to heavier, more complex equipment, with terms like backbone being used. Edge products tended to be used for less complex devices that allowed connection to the core.[72] To confuse things, a switch or router could be core or edge. To over-simplify—Cisco focused more on the core, 3Com more on the edge.

Primary Access, bought in March 1995 for $170 million in stock, was a remote access server company for network service providers such as telecommunications companies (or "telcos")—and another questionable acquisition. At this time, the 3Com salesforce was not aligned nor competent in selling into the telecommunications carrier space, which did not help 3Com's outcome. Executives said the acquisition was a classic case of when foregoing engineering due diligence leads to a bad result. What 3Com had bought turned out to be a Potemkin Village. (Perhaps the best-known Potemkin Village in modern memory is in a scene in the movie Blazing Saddles; the guys roll into town looking for Rock Ridge, only to discover it's just a wooden façade.) Primary Access's codebase was also a mess. Steve Russell, the VP of NSD engineering, worked hard to get the product moderately stabilized, and later became the VP for Primary Access.

3Com had dominated in the LAN space from the mid-1980s into the mid-1990s, but now the importance of the infrastructure for the Internet was becoming significant. 3Com's inattention to telcos, their product requirements, and selling methods that aligned with their expectations left it vulnerable; the decision to buy US Robotics later was an attempt to play catch up. Along with the prospect of the adapter business changing (and not for the better), one executive opined later that, "Eric was between the devil and the deep blue sea, and both of them wanted a serious piece of 3Com's ass."

---

72. For a more detailed explanation of "edge" vs. "core, visit 3comstory.com/tech.

The Nice Guys

While some within 3Com felt the best ATM move was to buy a company called Centillion Networks, (later bought by Bay Networks in 1995), the company moved in a different direction. Various executives noted that Eric prized the opportunity of getting technology from Israel. Eric recalled the movement behind what turned out a unique and somewhat revolutionary idea at the time:

> We made our first Israeli company acquisition in 1994, a company called NiceCom. We knew we needed to have at least two major backbone technologies in our portfolio. One of them was FDDI, and Synernetics helped us address that. But we also felt that FDDI was going to have a limited future, and that ATM may actually be the better technology for the rest of the '90s. We went shopping for an ATM company, because we had almost zero in-house expertise. John Boyle led that search. At the time, he was in Janice's group, focusing on M&A.
>
> We eventually selected NiceCom for its ATM technology. This was the first major technology acquisition of an Israeli company for the industry. NiceCom gave us access to the Israeli technology ecosystem, which, at the time, was not very well understood. It was nascent, but it was very promising. This gave us access to a very talented team of engineers led by Nachman Shelef, who subsequently became general manager of several divisions at 3Com, and became a good friend. [Nachman serves as an advisor to Eric's investment fund, and is also a limited partner and board member.] These relationships that were forged in 1994, they are going very strong 25 years later.

While Nachman had hopes to take NiceCom public in Israel, he settled for $60 million from 3Com. His company was essentially 45 engineers, one manufacturing guy, and Nachman. Eric, along with Nachman have used the NiceCom acquisition as a business case study at places like INSEAD in Europe, Stanford, and in Israel, providing analysis from the buyer and seller perspective.

Unfortunately, while successful in carrier sites, the more complex ATM backplane technology did not turn out to be the massive winner in the long run. ATM's vision to move to the user's desktop did not occur due to the emergence of cheap Fast Ethernet and Gigabit Ethernet LAN technologies.[73] After the advent of the Fast Ethernet standards, the FDDI and ATM backbones

---

73. For a more complete explanation of ATM and networking in the 1990s visit 3comstory.com/tech.

were less critical. 3Com's response with the CoreBuilder 3500 in 1997, that could run all three, may have been a case of too little too late.

For Israel as a whole, though, NiceCom turned out to be an enormous blessing. Receiving this validation from the first major Silicon Valley technology company opened up the floodgates to acquisitions and Israeli investment. If you track the acquisitions of Israeli companies, the first major one ever was 3Com's 1994 acquisition of NiceCom. While the $60 million price tag may not seem that big of a deal, it was a lot of money for a company that didn't even have revenues—the company was basically buying a product that was in beta test, along with 45 engineers. One acquisition of Israeli technology in 1994 grew to five in '95, and 20 in '96. This deal was a pivotal point that introduced the impact of Israel in the technology industry. After Eric left 3Com as CEO, he embarked on a number of initiatives that helped drive technology and employment in the country.

Bob Finocchio shared his memory of this period:

> We got an incredible team with Nachman Shelef. Eric was the engineer identifying what to buy, and I was helping execute all these deals. NiceCom was a wonderful experience for me. The team more or less got in business by being subcontract designers for somebody who was building a fiber optic hub. They had to meet a deadline to get this product out the door, and it was during one of the war periods when Tel Aviv was in lockdown. They finished the product in Nachman's apartment, when the family was wearing gas masks, missiles overhead, and it didn't matter that they might be attacked, they got the product out the door. Just wonderful, incredible, brave people.

Bob Finocchio also noted that Nachman and his crew were super smart. Many of them came from a military intelligence background, where they did digital signal processing that Israeli defense forces are very, very good at. Brilliant engineers who could get stuff done under pressure—nothing phased them. Avinoam Rubinstein, one of their top hardware engineers, came from a military specialty in defusing bombs.

Despite Eric and Bob's enthusiasm for NiceCom and its team, the war was being lost on one important front. As 3Com found value and talent in various global corners, Cisco was amassing a Silicon Valley juggernaut of talent via their first acquisition Crescendo, followed by Kalpana, then Grand Junction, that went on to dominate the Fast Ethernet switch market. The Cisco strategy to go local early with great networking talent all found in Silicon Valley, paid off in the end to be a great strategy, creating synergies across their product line, while 3Com systems foks dealt with plane trips and politics of globally dispersed divisions competing with each other. BICC Data Networks and Synernetics were great star acquisitions for 3Com, but ultimately faded in Cisco's planet.

"My Life Lesson Number 1—never bet against Ethernet."

Leon Woo, VP Engineering and Co-Founder, Synernetics

Chapter 12: 3Com's Cash Cow

Intel in Sight

Doug Spreng was a voracious reader growing up in Ohio—he built his own reading closet in his bedroom as a child. He worked hard at his boarding school to get into a good university like MIT, where he received his degree in Electrical Engineering, and added an MBA from Harvard for good measure. Following the good advice of others, Doug decided to pursue electronics in an era that John Young, president of HP from 1978 to 1992, dubbed the Golden Age of Electronics.

As one of the youngest general managers at HP, Doug rose through the ranks quickly from 1980 to 1990. His work focused on building products such as disk drives and then HP 3000 business computer systems. After leaving HP for a brief stint working for Paul Cook at Cellnet Data Systems (Cook had been the CEO and founder of Raychem Corporation, a $2 billion company sold to Tyco International in 1999), Spreng found his way to 3Com in early 1992. 3Com offered Doug the chance to use his HP experience without having to deal with the constraints of a large, slow-moving company like HP.

Doug recalled his approach to address Intel's incoming threat in the market for Ethernet adapter products. "The whole point was to say, 'Enough of this fear and trembling. We invented this technology and are the Master of It—we should attack the market aggressively and out-Intel Intel'". Doug noted that he had heard of a Japanese motorcycle manufacturer that had a "Crush Honda" campaign, saying they would crush them like a bug. Doug said, "It was audacious and aggressive and just what was needed to mobilize the Network Adapter Division against Intel. I rolled out our 'Crush Intel' campaign in the first week of my new job at our afternoon coffee meeting and it was designed to be multi-year."

3Com was always a very competitive company with a lot of talented people. It just needed to be led and organized, and Doug had experience doing that at HP. The Etherlink III was close to launch, and would prove to be a major asset in this battle with Intel. 3Com owed Andy Verhalen, the prior GM, and

Paul Sherer, who later became 3Com's CTO, big-time for this work. Doug said about Paul:

> I called him the DTO, the Division Technology Officer. He almost single-handedly invented some of the key elements. When you're already the market leader, and then you develop a superior product, especially if the cost structure is similar if not better, then it's going to be a license to steal. That's exactly what we did.
>
> Everything was premised on being aggressive. I've done a lot of reading in the past at HP and other places about how you take leadership progressively, in technology companies particularly. So, Geoffrey Moore, the author of *Inside the Tornado*, *Crossing the Chasm*, and all that was one of my heroes.[74] I started taking his ideas and insights and mapping them into semiconductors because that was where the big breakthroughs were happening.... in '94 or '95 perhaps, he was a guest speaker at our offsite, and the reading assignment was to read *Inside the Tornado* and understand the concepts.

The day before Doug joined, Eric had called him into the office and said, "Your VP of Manufacturing Jack Moses, just quit. But that's not necessarily a bad thing. You can hire whomever you want for this position." Doug described his own thinking, "'Is there anyone that I can go hire for this position? Yes! Randy Heffner!' I had worked with Randy in Boise, Idaho, for a while. I knew him as an innovative, very aggressive, implementer of operations."

Randy was VP of Manufacturing for Steve Jobs at NeXT at the time and had built a highly automated manufacturing line for NeXT computers. After Randy completed his commitment to Steve Jobs (described later on), he joined 3Com. NAD needed a great VP of Operations to drive down the cost and drive up the volume product. Randy had all the skills to do that. For one, he hired great people. He also could craft a clear vision of what the end goal would look like, and get everybody to accomplish that vision. Doug felt that the hiring of Randy Heffner was one of the most significant things that took place during his tenure. Doug's division fit nicely into Eric's GDN vision to provide adapters in the FDDI and ATM realm; also he had enough of an R&D budget to support all the adapter flavors the market demanded.

---

74. Geoffrey Moore also co-authored The Gorilla Game, and is a marketing and strategic consultant to high-technology companies and a VC at Mohr Davidow Ventures. Moore importantly described the model of when innovative products move from early adopters to larger market segments, or crossing the chasm (reaching the early majority), similar to the technology adoption lifecycle.

Doug also reflected on his experience around 100 Mbps Ethernet aka Fast Ethernet. Eric asked Doug to lead the project, assigning Ron Crane (who was now a consultant to 3Com) to the project. Doug's thinking was that, from his working with the author Geoffrey Moore:

> If you're a market leader, you lead a transition, you don't follow it. 3Com had led the Fast Ethernet consortium, including 3Com, National Semiconductor, Sun Microsystems, Grand Junction Networks, and LAN Media, Ron Crane's firm. Intel was not in the consortium. Most conservative companies follow the transition that allows new early challengers to come in and take part of your market away. We knew if we left that opening with Intel, they would take it. So, we jumped on the PCI bus (which Intel had invented as the PC standard interface) and did everything imaginable with the products we had with PCI. PCI was used to attach expansion cards in a computer. One of the reasons PCI came along in the first place was to provide a higher speed interface bus to handle 100 megabit Ethernet.

Ron worked with Paul Sherer on the project. Around the same time, HP attempted to install their own product, called 10Base-VG with the IEEE as the standard. Doug noted:

> When we saw HP's proposal, we said, "Holy shit. These guys are gonna preempt us." Because they were on the IEEE committee, and there were other companies that were friends of HP on the committee that might vote their way. We had the opportunity to comment and propose and so forth. Without knowing exactly how, we went to IEEE at the next meeting, and proposed to do a Fast Ethernet using category 3 wiring technology. Hewlett-Packard's proposal was proprietary; it wasn't the standard at all. At any rate, it turned out... our proposal was accepted. But keep in mind that we still didn't know exactly how to do it yet—this wasn't based on Ron Crane's technology. We basically told Paul, another engineer named J.R. Rivers, and his merry men, "Guys, you gotta do it." And I think at about one IEEE meeting, three months later, they were able to demonstrate the technology.
>
> 3Com went on to be the market leader in Ethernet adapter products, and went into an enormous upgrade cycle, which is always incredibly profitable. Hiring Randy was one huge win, and the Fast Ethernet

> story was another one. I'm proud to have helped lead this and it did have enormous impact on the company in a positive way.

When Kevin Canty was hired by Randy Heffner in 1992 into what was called Personal Connectivity Operations, or PCOps, (largely, network interface cards known as NICs aka network adapters), it was a great opportunity to complete what he and Randy had begun at NeXT Computer under Steve Jobs. With discrete designs for NIC cards of various flavors, 3Com had hedged its bets with Ethernet and Token Ring. Token Ring had certain advantages, but didn't scale well. Doug Spreng, Jim Basiji, and Randy Heffner made the key decision to build their technology on ASIC silicon, not placing discrete parts on fiberglass (printed circuit cards). This PC card process was costly and margins were poor; ASICs offered them a way out. As David Packard said, according to Kevin Canty, "The only thing worse than a shitty business is a big shitty business."

3Com developed the silicon relationship with AT&T Microelectronics, later part of Lucent. This decision built up the margins of the PC card business from 20+ percent to 50+ percent, which helped to generate the cash flow needed to fund 3Com's systems business and to strengthen the company. And with the launch of the Etherlink III products in 1992, 3Com went head-to-head with Intel.

NeXT Steps

Randy Heffner now runs a 2,400-acre cattle ranch in Idaho. Randy runs roughly 300 pairs (a cow and her calf) and also grows pasture alfalfa and grass hay. Randy recalled his pre-3Com days with Steve Jobs at NeXT, "Working for Steve for five years was, at the end, a net positive. The highs were so great, but the lows were so crummy." When he interviewed with Steve for the job, he remembers it as a hardcore grilling—to call the interview a "debate" would be too polite. Randy also recalls almost being fired by Jobs for having a transportation van that was the wrong shade of white.

During this time, Randy also had to please Ross Perot, the founder of Electronic Data Systems (EDS) and two-time third-party presidential candidate (in 1992 and 1996) known for his straight-shooting style, along with Ross' friend and CFO, Thomas Walter. EDS was an investor in NeXT, and Randy recalled that "Ross and Tom were very frank, firm, asking me 'Randy, what the hell's going on here? How come you're not making your numbers? Got a problem in the factory? What? You got to tell me what the problems are, right now. You need to list those out right now.' This little 'chat' went on for over an hour."

After this experience, Randy figured he could survive just about any difficulty in business. Randy later had a falling out with Steve and left NeXT. Steve was afraid Randy would take his skills to HP or Sun, so he paid Randy to not work for six months. It was after this that Doug was then able to take Randy into 3Com. Randy noted, "I always wanted the manufacturing engineers to be every bit as good as the R&D design engineers." This was a key reason why Randy and his manufacturing team were so amazingly successful. They placed great importance on the design and throughput of the factories they built in Santa Clara, and later Ireland and Singapore.

Jim Basiji, the head of R&D, and Randy worked together to make sure their products were designed for reliability and manufacturability. Greg Arnold, another NeXT alum, helped champion the testing. Randy wanted to insert the test engineering process, including working with R&D to include design for testability, into the production line, working to achieve line balance, then increase the bucket rate per product to increase throughput, reduce cost, while maintaining quality.

Doug's team was to be the high-margin, high-efficiency cash cow that would acquire and nurture the Network Adapter Division business, now called PCOps. Jim Basiji, VP of Engineering, and Kirk Blattman, Director of Engineering, helped lead the ASIC design work. Based on his contacts from NeXT and knowledge of their factory, Randy hired NeXT factory automation, test automation, and materials workers. Some of those who were instrumental in building up the great factories that helped 3Com succeed were Greg Arnold along with Nick Mitchell from NeXT, Jack Gilbert who ran automation engineering (instrumental in the production line), Larry Nunes who ran the production for Santa Clara, and Kirby Hansen who oversaw Manufacturing Engineering.

Over time, this team was able to reduce the production time for a NIC card "bucket rate" from 40 seconds down to two seconds. This made 3Com globally competitive, without the need to outsource. Because of its success, Donal Connell had been hired in Ireland to replicate the NIC factory there.

This automation effort represented a cutting edge, world-class process. Even before so much influence from the NeXT graduates, 3Com had been independently innovating along similar lines, going back in its product generations. According to Dan Robertson, who oversaw much of 3Com manufacturing's growth between 1983 and 1989, "3Com had been planting the seeds and developing the skills, workforce, production technology, and capital equipment investment that would eventually support Randy and Kevin's success with PCOps." 3Com's production processes were first developed and refined by Gary Heidenreich, Jim Hayes, and Dan himself. When 3Com's expanded production lines were set up at the Great America campus, Jack Moses and Larry Nunes carried its manufacturing into the next generation by updating automated processes developed at the Kifer Road site in the late 1980s.

But finally, it was Kevin and Randy who refined the manufacturing process again in 1992, this time for worldwide expansion. They brought the operations to the next level of design-for-manufacturability through the crucial integration of both test and process engineers with product designers. Having these teams work in close quarters opened the gates to faster and better innovation. The resulting manufacturing process demonstrated that outsourced and subcontracted manufacturing were not necessary if automated processes, skilled training, and top engineering were put in place with high volume lines pushing out millions of adapters per month.

Manufacture On

The critical secret of the manufacturing process that Randy and his team brought to 3Com was that coupling manufacturing with the engineers designing the products enabled the company to out-innovate its competition by implementing improvements on a continuous basis. A high level of automation and well-designed assembly line helped 3Com maintain lower costs and higher margins for many years than its competitors manufacturing in Asia, who also had to deal with shipping costs, import duties, and other factors that drove their prices up and their margins down. Kevin noted, "3Com engineers would go down to the floor and work on the placement of parts for optimizing time, cost, and quality" of the products. In addition, in 1992 3Com received the President's Environment and Conservations Challenge Award for its ground-breaking CFC-free manufacturing.

3Com could pour out millions of NIC cards per month in a process that was efficient and optimized by its manufacturing strategy—one that also supported the tax strategies implemented by VP of Worldwide Taxation, Eileen Landauer. The company could still get a manufacturing cost structure in offshore, tax-treaty favorable countries, such as Ireland and Singapore. During the high-profit years of the 1990s, 3Com saved millions of dollars by cutting its effective tax rate. With the Tax Department's offshore tax strategy, 3Com used the Cayman Islands entity offshore cash to reinvest in projects such as the Singapore factory.

PCOps Manufacturing had a work hard, play hard ethos. There would be some turf battles, and the occasional love/hate relationship between Doug Spreng and Bob Finocchio. Notably, Doug's PCOps group also had some hub offerings, and they felt that BICC's hub products would be a major growth catalyst for PCOps' revenue line after its purchase in 1992. Instead, the hub and switch management oversight and revenue credit were given to NSOps (Bob Finocchio's Group), and PCOps became a supplier to NSOps via their factories in Ireland, Singapore, and Santa Clara. BICC became known as the

Premises Distribution Division's (PDD) design team, led by Edgar Masri at the time, a key division within NSOps. Arguably, the stackable hub and switch products that came out of PDD were the second most significant revenue generators, with adapters as number one.

Contributing to this success was a laid off DEC engineering team in Ireland that Ron Sege had found after the BICC acquisition, which gained 3Com key ASIC technology development talent. This Ireland team was led by Tadhg Creedon. Per Tadhg:

> We were tasked with 3Com's first low-cost Ethernet switch ASICs. Our mentors were John Hart and Paul Sherer. John helped us by creating an almost mathematical definition of a switch which we constantly checked to ensure correct behavior, while Paul encouraged us to take risks in minimizing buffering requirements and other features from his work in the highly-successful adapter world, such as PACE (priority access control for Ethernet).
>
> The result was a very low cost per port Ethernet switch which ultimately replaced hubs, as the market demanded, allowing 3Com's stackable products to become the world's leading low-cost switch and a very profitable business for 3Com. PDD's Steve Carter and Terry Lockyer referred to it as 'switching for the price of repeating'.

Bob Finocchio had this to say about his sometimes contentious relationship with Doug Spreng:

> At some point during all of that, Eric made me president of 3Com Systems, (the old NSOps) and I took over the salesforce as well as the products. Doug Spreng ran adapters. The adapter business was printing money, while my side of things was a break-even business. But I believed the adapter business was going to go away.
>
> The first big blow was when Intel decided it was going into the adapter business, and eventually made its own silicon. It was easy to see that the silicon was going to go in the motherboard of the PC. People were not going to buy cards anymore. And at the time, Intel was a huge OEM of PC motherboards. They were in the systems business; it was a multibillion-dollar business for them. So, it was clear to me that the adapter business was going to go away. Toward the end of my tenure, I wanted us to double-down on systems. It was clear we

> had a lot of rationalization to do in terms of our technologies. We weren't spending R&D 100 percent efficiently because we had so many different platforms. And Doug wanted to double-down on the edge of the network, because we were making all the money there.
>
> I respect his expertise, he's a great product and manufacturing executive, and he'd built out effective low-cost, two-tier distribution. My side of the business was going more and more direct, because they involved some complex system sales. So, there was a divergence.

Notably, Finocchio's prediction was correct in his assessment of the adapters' demise, although not precisely with the right timeframe, as 3Com did continue successfully with adapters until the late 1990s. But this pivotal risk was not confronted and dealt with soon enough, resulting in the complete collapse of the product line after Bruce Claflin took over in the 2000s.

Chapter 13: A Cult Culture

People generally start to hold individuals accountable once they reach their teenage years. We're no longer so quick to forgive or dismiss destructive behavior as mere growing pains or as the actions of children who don't know better. Our attitude towards the behavior of companies follows a similar trajectory.

We tend to forgive the quirky—and sometimes toxic—culture and behavior of young startups. But as companies mature, especially once they've gone public, our expectations become far more stringent. With people, we examine their character, moral fiber, and internal compass; with companies, we look at their corporate culture, mission and vision statements, and tone from the top. As 3Com lands firmly in young adulthood, we'll pause the chronology for a moment to take stock of 3Com's culture—on a range of topics including the esprit de corps, financial integrity, gender equality, the employee code of conduct, education, patents, and real estate (an unsung piece of the corporate culture puzzle).

Corporate Camaraderie

Andy Verhalen recalls that 3Com in the late 1980s was simply a great place to work. And, "the reason it was a terrific place to work was because of Howard Charney. People loved Howard. He was the spirit of the company. He was the glue that kept people wanting to come to work every day." During

Howard's time at 3Com, Andy recalls that there was a sense of openness and transparency at the company, especially at Friday meetings with all employees. "Anybody could ask any question." Even the kind of numbers and performance metrics that are usually only known to the top executives were shared not just with all employees at the Friday meetings, but also with guests and visitors who might happen to be in attendance. According to Andy, "When Howard left, frankly, it was a sad day for the company.... I don't think the company was ever quite the same, culturally."

And sometimes the benefits of 3Com's culture paid off in surprising ways. In 1990, 3Com had a sizable layoff as a result of exiting their agreement with Microsoft. The company worked hard to manage this unfortunate situation as best they could. A year later, 3Com hoped to hire back some of the people it had laid off. One engineer who had gone to work at Seeq, got a call from 3Com and told his current manager, "My company has called me back. I have to go." Even in the low points, folks knew what the company was and what it stood for. For many people, 3Com would remain "my company," even after it had laid them off.

Financial Culture

The tone emanating from the top was constructive and ethical, and with a growing demand for its products, there was only occasional pressure to "make the financial numbers" at the end of challenging quarters. CFO Chris Paisley ran a very tight ship, with great ethics. Despite that, in some years there were the usual pressures to smooth out revenues by either "pushing out"—that is, delaying product from shipping—or sometimes the more questionable tactic of "pulling in" shipments from the next quarter in an attempt to increase revenues in the current quarter. Companies like Cabletron appeared to be masters at it, showing quarter on quarter consecutive growth rather magically.

3Com sold to distribution channels and recognized their revenues on "sell in" to the channel, while simultaneously booking sales return reserves that reduced revenue to attempt to match "sell through." The net effect would be to only recognize revenue based on what was actually reaching the customers factoring in estimated future product returns. While this could provide a convenient way for meeting financial figures at end of quarter, 3Com's accounting was viewed as ethical and honest. Per Cindy Hawkins, 3Com's Corporate Controller (later Network Systems Operations Group Controller), "I saw first-hand the 'do the right thing' ethics from our early leaders and those who succeeded them. This mindset was shared by the high caliber people throughout the finance organizations, many with CPA backgrounds. I always had full confidence in the integrity of 3Com's financial reporting."

Echoing this point, Mary Henry, a Managing Director of Goldman Sachs who followed 3Com during its late 1980s and into the 1990s, said, "Chris Paisley remains one of the iconic technology CFOs—intelligent, high ethical standards—he was a great asset to 3Com." Indeed, as mentioned earlier, Chris's ethical and smart leadership was part of 3Com's success, both internally and externally, in helping Eric set the proper "tone at the top" and in dealing with Wall Street.

Gender Supportive Culture

Marina Levinson also agreed that 3Com's corporate culture was ethical and positive. From 1987 to 1999, she held leadership roles in the management of information technology for the company. In that capacity, she was also a leader in 3Com's global integration. As an immigrant from Russia, Marina arrived in the U.S. with a degree and not much else. She has gone on to serve as the Chief Information Officer for both Palm and NetApp, and she has worked as an advisor to Andressen Horowitz and Silicon Valley Bank, among others. She is currently a partner at Benhamou Global Ventures.

Marina recalled how strong and fair 3Com's culture was under Bill's, and subsequently, Eric's leadership. "It's incredible how young the company was, how small the company was, yet it had a very good definition of its culture, who they were, what the values were, with pay for performance and meritocracy. I was able to recognize that and appreciate that only much later on in my career when I joined companies that were bigger but not as mature."

Marina also described her experience as a woman within 3Com's ranks, "The company was incredibly female-executive friendly, and in fact with all the discussion about sexual harassment and discrimination of women in tech, I sincerely can say that I never felt disadvantaged by being a woman." In fact, she says, "I was able to do everything and anything that I set my mind to." She attributed the lack of sexism within 3Com largely to its culture of meritocracy. When she brought up the favorable treatment of women at 3Com with Eric, she remembers his response being somewhere along the lines of, "But of course, meritocracy wouldn't care who you were."

The combination of good management training, plus the tone at the top culture of strong ethics and moral behavior, was first set by Bill Krause, who brought it from HP, followed by Eric, and reinforced by Debra Engel as VP of HR. It remained throughout 3Com's history. Debra Engel recalled a moment when "I was in a customer meeting with some of my fellow executives and the customer made inappropriate comments to me. Bob Metcalfe was within hearing distance, came over to the customer and told the customer that the comments/behavior were not appropriate, and he could either apologize or we didn't need their business." Debra was 31 when she was made VP of HR at

3Com, and no one focused on her age, just her performance. She also noted in her 15 years at 3Com, they only had one HR-related lawsuit, which they won, because they had "done the right thing."

The Six O'Clock News Culture

A code of conduct helps set the proper culture. And in fact, there was virtually no litigation related to 3Com breaking agreements, violating laws or regulations, defective products, or causing harm—a testament to a strong culture of "do the right thing."

Mark Michael shared an ironic story about 3Com's formal adoption of a code of conduct back in 1989. The company's auditors at Deloitte had noted that 3Com had not yet formalized any such code, and Chris Paisley asked Mark to write one up, which he did. He presented a draft to Chris and Bill Krause. They were in Bill's cubicle discussing the section related to business gifts—specifically, a rule against accepting excessively generous gifts in various forms—and the common-sense test for what would violate the "6 O'Clock News Rule": never accept something if you wouldn't be comfortable having that fact shared on the 6 o'clock news.

The phone rang, and Bill's admin said that Chiat Day, one of the big advertising agencies competing for 3Com's business, was on the line to offer Bill a World Series ticket for what would become the famous third game in the 1989 Giants vs. A's matchup. He looked at Mark and Chris and said, "I guess this wouldn't pass our new policy about undue influence and gifts." Heads nodded in the affirmative and Bill declined the invitation, at which point Chris excused himself and said "I gotta go." This World Series game happened to fall on the day of the 1989 Loma Prieta earthquake. It also so happened that Chris was sitting in the front row as a guest of none other than Deloitte when the quake hit—Deloitte being the firm that had prompted 3Com to adopt the code of conduct which questioned such gifts in the first place! Because cameras were rolling at the game, it became the first major earthquake to be broadcast live. News crews started taking video of the fans in the front row seats, and Chris actually ducked any chance of an interview so he wouldn't be on the 6 o'clock news!

Implementing the code of conduct in 3Com's operations around the world, Mark reflected, was a useful, preemptive way to stop misconduct from rising to the top. Outside the U.S., 3Com employees were always curious about how serious and sincere the company was about this stuff. An ethics hotline was established, though rarely used. Over time, Mark came to believe that employees wanted to handle sensitive issues locally, rather than escalate them to regional management, and certainly not to HQ. This was especially true for locations outside the U.S. Going forward, only matters that couldn't be successfully addressed locally were brought to the attention of senior management.

A Classroom Culture

NetDay activities (left to right standing behind the students): Tom Payzant, Harvard professor and author, and former Superintendent, Boston Public Schools; Leon Woo, Co-Founder Synernetics; Tom Menino, Boston's longest serving mayor from 1993 to 2014. Photo taken April 1997.

3Com led efforts to help schools and communities gain access to the Internet. On March 9, 1996, 3Com and others launched NetDay with Al Gore and Bill Clinton. 20,000 volunteers donated time to wire 20 percent of California's schools, with millions in donations of equipment over the next decade. NetDay continued annually until 2006, when it merged with Project Tomorrow. David Katz, 3Com's VP of Strategic Alliances, helped on numerous school projects in California and later worked with the Mayor of Boston to network schools in the Boston area, winning the U.S. Conference of Mayor's Award for the best public-private partnership in 2000. This began 3Com's Urban Challenge, which expanded to 60 cities by 2006.

David Katz remembered how supportive Eric was of these endeavors, "I remember Eric saying that there was value in learning how to network schools whether or not the market ever developed." Participating in NetDay allowed 3Com to prepare for the inevitable demand of connecting students and teachers to the Internet. As the market developed, 3Com was positioned as the leader and ready to take advantage of the opportunity. With the help of Debra Engel, 3Com also began donating older-generation products to K-12 schools across the country. David said, "It grew organically and the company's expertise grew organically as well. Donations and commitments were made and it led to a mutually beneficial outcome for all parties involved."

In Silicon Valley, 3Com networked Independence High School in the East Side Union School District, a school with over 3,000 students. David recalled:

> It's this massive campus, and we got the whole thing networked in one day. There were probably 200 people from 3Com. You talk about a soul of 3Com—that day was incredible. It left the infrastructure

> that jumpstarted these schools to get connected in a way that really transformed education.

The education sector became one of 3Com's biggest markets. As the head of marketing for education, David said, "There was a passion in the company for getting these schools connected. We became one of the largest players in the K-12 school market. 3Com networked huge segments of schools in Silicon Valley, Southern California, Delaware, Utah, and even globally in Australia, New Zealand, and Europe. There really was this spirit of giving as it unfolded."

3Com was also involved in supporting education at the university level. In 1998, Eric and David helped launch Western Governor's University, which remains one of the most prominent virtual universities. 3Com was also a founding member and contributor to the Stanford Network Research Center, launched in 2000. Other research funding was provided to U.C. San Diego, National University of Singapore, and Carnegie Mellon. In 1999, Greg Shaw, along with Bob Metcalfe, 3Com, and ten other 3Com and MIT graduates funded a $2 million endowment at MIT to support a full professorship in networking technologies. The first holder of the 3Com Founders Chair was Tim Berners-Lee, the inventor of the World Wide Web. Bob Metcalfe has also personally endowed a professorship at MIT to support writing, engineering, and entrepreneurship.

A Patent Culture

Many people who work in tech don't often recognize the impact that a really good legal team can have on their company. The role of a legal team isn't just to minimize lawsuits and prosecution, or maximize contractual negotiations. A legal team can also contribute significantly to the bottom line, and to a culture that values innovation.

During the "crossing the desert" period in the early 1990s, Mark Michael and his legal team wanted to create an intellectual property portfolio that would generate revenue and provide a defensive posture against competitors. Early on, opportunistic competitors began vying for a piece of 3Com's hide via license revenues for patent violations. For example, Mark was invited to a meeting with AT&T, which had studied 3Com's products and technology and concluded that 3Com needed a license. AT&T's position was, "We have many tens of thousands of patents, and you surely must be violating something—we'll offer you an across-the-board license if you pay us 5 percent of your revenues." Mark dutifully presented that "offer" to his management, which declined it. IBM similarly wanted to own 5 percent of 3Com's hide. Other large companies—including Motorola and Intel—continued knocking on the door over the next two decades.

3Com had started with an empty patent portfolio. Ron Crane's Ethernet patent for a "local computer network transceiver"—filed in 1981, and granted in 1984—got the ball rolling. John Hart, 3Com's Chief Technology Officer (CTO) worked with the legal team to create an incentive program to encourage engineers to file patent applications and build out the intellectual property portfolio.

The program was an engineer's "President's Club" Patent Award Banquet, similar to the types of programs sales departments often use to motivate top performers. The engineers would come up with patents—using John's budget for the filings—and would be awarded points for patent filings and for issuances. Earning a certain number of points earned you a ticket to an enticing destination—first nearby Carmel, then later to more exotic locations including Portugal, British Columbia, Hawaii, and Martha's Vineyard. Guest speakers included Marc Andreessen, Vint Cerf, and MIT professor Michael Hawley.

John Hart shared a story about how he knew it was working well:

> This woman came up to me at the end of one of the early conferences and said "You know you saved our marriage." And that wasn't what I had expected. I looked at her and she said, "You know, we never get to go anywhere. My husband has his head in technology. It's the most boring thing but we have these conferences and these things are just the most fun and they're wonderful." And I was thinking—I could imagine the pressure on this engineer, are you going to qualify? It's a huge competition to get to the inventors' award night. And they kept getting better.

By the early 2000s, 3Com had built an industry-leading portfolio, with 1,200 patents filed and 800 pending. This gave the company great design freedom, without having to fret about the costs of licensing. Indeed, until after the acquisition of USR (which did have a licensing arrangement with IBM) and after Bruce Claflin came on board, the company paid no royalties. According to Mark Michael, "Our patent portfolio was more extensive than any of our direct competition, including Cisco. We did not initially seek to monetize the portfolio, although in skirmishes with Bay Networks and Intel, we went to court and collected single digit millions, royalty free cross licenses, or covenants not to sue for a period of years from them."

Mark Michael kept kicking the royalty license can down the road with AT&T, and 3Com never reached a final resolution that might have required the company to pay them a license—which qualifies as a resounding success. However, 3Com began to lose its leverage with IBM after the USR deal, since USR had a license with IBM. 3Com had never had a licensing arrangement with IBM, but during Claflin's tenure, it did begin to pay royalties. Nonetheless, even when the company was hemorrhaging cash and incurring product losses across every category, 3Com's patent license revenue remained profitable.

Real Estate Sets the Mood

In June of 1999, according to Abe Darwish, VP of Real Estate and Site Services, 3Com owned 5.4 million square feet of real estate and leased another 1.3 million square feet, covering 182 locations in forty-five countries for 13,000 people. (These numbers would shrink dramatically during 3Com's next chapter.) Over his years working on the first Kifer Road (Santa Clara) site in 1986 to building up 3Com's global sites, Abe always felt that real estate wasn't just about providing a place to park workers—it played a major role in nurturing culture and morale.

Abe always focused on creating and maintaining five elements for the ideal work environment:

- Open, light, airy, fun, and colorful.
- Embodying understated elegance.
- Unique, dynamic, and stimulating, yet quiet.
- Egalitarian, functional, and not based on status.
- Showcasing products in the best light.

"I remember coming to Marlborough, MA for my interview and walking into the lobby and seeing this beautiful life-sized painting of Jimi Hendrix in the lobby. At the time, as a guitar player of 20 some odd years I chalked it up to a nice touch. A familiar feeling of appreciation for art. But in reality, it came to represent innovation in the tech industry. A 3Com attorney once told me how many patents the company held and I was astounded. God I miss those days and all the wonderfully gifted people that worked there. I have truly not had better."

—Marc Wright, Desktop Support, 2008-2010

Chapter 14: The Middle '90s

Hitting One Outta the Park

While 3Com's reputation was on the rise with its strong financial and product performance, its name recognition remained limited. That was all about to change—at least among sports fans. The naming rights to the San Francisco stadium formerly known as Candlestick Park were licensed to 3Com from September 1995 until 2002, for $900,000 a year. During that time, the park became known as "3Com Park at Candlestick Point" or, more commonly, "3Com Park." (The stadium was still lovingly—or resentfully—referred to as "The Stick" by many locals and die-hard fans.)

At the time of the deal, newly constructed parks were frequently being named for businesses, but it was quite unusual for a company to go after the naming rights of an existing park. In the case of renaming Candlestick, 3Com's window of opportunity arose because San Francisco had recently won the opportunity to host a Super Bowl. Because of this impending milestone, the city needed to raise significant funds to upgrade the stadium, so it partnered with the 49ers to present a deal to potential companies.

Chris Paisley became 3Com's spokesperson for the deal. It may seem odd for the CFO, rather than the CEO or an executive in PR or marketing to be the public face for something like this, but Chris had the advantage of being an avid sports fan. Eventually, of all the companies that the city and the 49ers had approached, 3Com's offer was accepted. 3Com would pay out $4.5 million for a five-year contract, rather cheap by today's standards. But these initial negotiations left out the other team that played at Candlestick Park: The San Francisco Giants. Apparently, the Giants were still intent on calling the park Candlestick—unless a separate deal could be reached. When Chris found out, he remembers telling Eric, "The Giants are here with their hand out and if we don't allow the Giants to participate in this, you're going to have 81 baseball games a year where it's called Candlestick and only 8, maybe a few more playoff games of football, where it's called 3Com Park." Chris got the go-ahead to pursue a deal with the Giants from Eric, but that deal didn't materialize until the very last minute (literally). Within an hour of the live press conference announcing the name change to the park, Chris managed to close a deal with Larry Baer, the Giants' COO.

After the renaming deals were complete, Chris still had to face the potential public backlash over corporatizing a landmark. He could foresee the controversy that would arise from renaming a beloved park like Candlestick. He was asked to appear on the local CBS late night news to explain 3Com's perspective. Chris knew that the interview with sports anchor Dan Fouts could potentially get a bit hostile. Luckily, Chris had the near-midnight slot, which would get considerably less attention. As he was getting prepped to go on live TV, Chris turns to Dan and says, "'Dan, you never told me what you're going to ask me?'" Dan responds, "'Why? That's what I'm going to ask you. Why? Why? Why?'"

Chris remembers the "real vehemence and challenge in his voice." Despite this, Chris felt relatively steady as the interview began. As David Abramson (who deserves the credit for the original idea of 3Com Park) told Chris on the way to the interview, "When in doubt, act like a politician"—meaning as long as Chris keeps talking, he doesn't necessarily need to answer Dan's specific questions. Looking back, Chris remembers with much amusement that Dan's opening question for him was, "'How does it feel to be the most hated person in the Bay Area?'" The interview was indeed hostile, but Chris held his own. At the end, Dan even approached him and gave him kudos on representing his company well.

Looking back, Chris said that he had not fully estimated the impact that the Candlestick deal would have on 3Com's success. "It turned out that our stock rose significantly in the next two weeks, probably up 15-20 percent, and 3Com was in the news every day. The number of applications we got for job applications was probably three or four times higher than what we usually got. 3Com got a ton of prominence out of this. There were articles in papers across the country." While the name remained controversial, it did offer the company great visibility. Chris remembers once exiting an airport and seeing a limo driver who was waiting with a sign for a different 3Com employee. "The guy in front of me sees this guy with the 3Com sign and he leans over to him and says, 'It's Candlestick, damn it!'"

When the deal and drama were done, the renaming ultimately turned out to be a solid base hit (or perhaps a home run). Though some locals remained grumpy over the name change, 3Com was only the first of many companies to hold the rights to Candlestick Park. And if things had played out just a hair differently, there might have been a chance that 3Com could have gotten its name on the park for nothing at all. When imagining names for the company, Bob Metcalfe said in retrospect that he wished that he had thought of names that would convey radiance, or some kind of "beacon of light." "If I had the foresight," he said, "I think I would've just called the company Candlestick."[75]

---

75. As this book was going to press in 2019, it was announced that Oracle was awarded naming rights to the San Francisco Giants Stadium, replacing AT&T, for more than $200 million over 20 years.

What's in Your Wallet?

In 1995, Bob Metcalfe famously predicted that the Internet was growing so quickly that it was bound to collapse within a year. He even promised to eat his words if he was wrong. When the collapse didn't happen, he lived up to that promise by pureeing a printed copy of his InfoWorld column in a blender, and sipping down the mashed words.

> **"I predict the Internet will soon go spectacularly supernova and in 1996 catastrophically collapse."**
>
> —*Bob Metcalfe, 1995*

In retrospect, it's pretty easy to see where his skepticism was coming from. The growth and the pace of market disruptions were bordering on fantasy. Even in the 3Com boardroom, the news was almost too good to believe. A global data network was indeed emerging, and 3Com seemed poised to lead it.

A financial snapshot of the company in the mid-1990s painted (accurately) a pretty rosy picture. In March 1994, Chris Paisley presented the board with great news: The company had just closed on two useful acquisitions—Synernetics, with advanced switching, and Centrum Communications, a play in personal routing. Synernetics was now called the Switching Division, and Centrum had become the Personal Office Division. Q3 was their best quarter to date, with Q3 Sales of $219 million. Annualized sales per employee doubled from $209,000 per quarter, to $412,000. Adapter sales were $121 million or 55 percent of their sales. Network Systems business was strong on the back of the new NetBuilder and boundary routing orders, and customers DEC, Wells Fargo, ICL, and Anixter were their larger customers. In the BICC part of the systems business, channels such as Anixter, Ingram, and Tech Data were selling lots of stackables. Gross margins had climbed from the mid-40s up to 52 percent in Q3 1993.[76]

Sales continued to expand rapidly from $219 million in Q3 FY 1994 to $430 million by Q3 FY 1995, an almost 100 percent increase year over year. Synernetics and Centrum sales contributed, along with contributions from the NiceCom acquisition in 1994. In May 1995, 3Com also purchased a UK-based ISDN company called Sonix Communications Ltd.

At this point, the global data networking market had become a three-horse race among 3Com, Cisco, and Bay Networks (the $2.7 billion stock combination in October 1994 of Wellfleet from Billerica, MA, and SynOptics from Silicon Valley).

---

76. Bloomberg News. "3Com Agrees to Purchase Chipcom for 700 Million". New York Times. July 1995.

In September 1995, Chris Paisley was again in front of the board, reporting that 3Com had executed a fantastic Q1 FY 1996 (quarter ending August 30):

- The company had had a 2-for-1 stock split on August 25, 1995.
- For the first time, systems sales contributed 52 percent of the company's record quarterly sales of $430 million. This number showed the strength of the Premises Distribution Division's stackable hub known as SuperStack, their LinkSwitch products, and the Switching Division's Lanplex product line.
- 3Com's gross margins were 54 percent (with operating expenses at 32 percent, dropping a solid 21 percent in operating margins).
- Headcount was 3,427. Revenue per employee was $519,000, a very good sign of productivity.[77]

Leading up to the USR acquisition in 1997, 3Com's financial picture was still looking pretty solid, with good margins overall. The financial news was strong during these boom years. FY 1996 Sales were $2.3 billion, with an operating income of $424 million, or roughly 18 percent. Net income was $280 million, or 12 percent.

It's illuminating (if a bit accounting wonkish) to consider the value of 3Com's adapter business by looking at the profitability by division below. As Wall Street had speculated for many years, the adapter business was 3Com's cash cow, and provided a solid operating margin of 30+ percent to the company, effectively funding investments in the Network Systems Operations Group, which was running at only a 13 percent margin. And the theme for the 1997 Operating Plan was all about growth; size was clearly on the minds of the company's leaders.

P&L by Business Lines, FY 1996, one year prior to USR Acquisition

P&L Owner:	Doug Spreng	Bob Finocchio						Richard Joyce	Alan Kessler	
		Network Systems Operations								
Business Unit:	Personal Connectivity Operations	LAN Operations	Wide Area Network Operations	Premises Distribution Division	Network Management Division	NS Ops Management (cost allocations)	NS Ops Total	Remote Access Products	Customer Service Operations and Other	**Total**
Products:	*Network Interface: Ethernet, Token Ring, FDDI, Mobile*	*LAN products: CellPlex, Lanplex, Linkswitch, etc.*	*WAN products: Netbuilder, Primary Access Servers, etc.*	*SuperStack hubs, etc.*						
Net Sales	$958	$511	$306	$461	$15		$1,293	$31	$45	**$2,327**
Operating Income	$295	$22	$22	$145	($2)	($23)	$164	($12)	($23)	**$424**
Operating Income %	31%	4%	7%	31%	-13%		13%	-39%	-51%	**18%**

Source: 3Com FY 1997 Operating Plan Document, FY 1996 Actuals Section

77. Paisley, Chris. "Internal memo: Q196 Summary of Operations". September 1995.

The 3Com board that Chris was addressing, by the way, was an amazing crew with an abundance of intellect and experience. Here is a brief introduction to some of the board members in 1997:

- Gordon Campbell—founder of Seeq Technology (the company that provided 3Com the valuable first working VLSI chip for the Etherlink product), Chips and Technologies, and 3Dfx Interactive. In 2000, he went on to found Techfarm Ventures.
- Jim Barksdale—Netscape's leader until its acquisition by America Online. He then founded the Barksdale Group, an angel investment firm, and co-chaired George W. Bush's Information Technology Advisory Council. Jim's ideas and support were instrumental in pushing Eric into the CEO role in 1991.
- Dave Dorman—employee 55 at Sprint Communications in 1981, CEO of Pacific Bell in 1994 at age 39. In 2002, Dave was CEO of AT&T.
- Jean-Louis Gassée—executive at Apple Computer from 1981 to 1990, later founded Be Inc., Chairman of PalmSource, Inc. in 2004.
- Philip Kantz—CEO and President of TAB Products Co., Trans Ocean Ltd.
- Steve Johnson—Co-Founder of Komag Inc., a thin film media innovator, and CEO and President from 1983 to 1999. Steve was astute, ethical, and supportive to 3Com. Sadly, Steve passed in March of 2018.
- Bill Zuendt—President of Wells Fargo until 1997, Bill led Wells Fargo's push to be an early leader in using technology in the stodgy banking world. He was an important and vocal member of the board.

Irrational Exuberance

Life seemed good. It was true that Cisco's sales were edging ahead, but 3Com had a new strategy, a quiver full of new talent and new products, and a vision to get its mojo back. And while 3Com was in the midst of an acquisition binge, the company itself was the occasional target of acquisition discussions. There was a time in 3Com's history where Compaq considered buying the company. Eric Benhamou had entertained discussions with Eckhard Pfeiffer, the CEO of Compaq between 1991 to 1998, during the hyperbolic boom years for the PC market. (After ultimately buying DEC in 1998, Compaq began losing market share to Dell and was sold to HP for $25 billion in 2002.)

Even as late as 1993, 3Com had bigger sales than Cisco ($723 million vs. Cisco's $649 million). The trendlines were unpleasant, but the gap was still small enough that it could be closed with a couple of hit products from the company that had pioneered small business networking solutions.

By 1996, 3Com booked $2.3 billion in sales, while Cisco's sales were almost twice that at $4.1 billion.

And by 1997, 3Com's revenues had risen more than 13-fold, from $419 million when Eric became CEO in FY 1990, to $5.6 billion after the USR acquisition. In May 1997, 3Com had a market capitalization of $16.2 billion. However, while 3Com may have finally emerged from the desert, Cisco had taken advantage of its position in the enterprise market, its product strategies, and in execution. Cisco's market cap was a whopping $53.5 billion in July 1997. And by 2001, Cisco was nine times bigger in sales ($22.3 billion vs. $2.4 billion).[78]

There was no questioning 3Com's impressive growth. There was also no questioning that Cisco had dramatically outpaced 3Com. There was really only one question that mattered in the long run: was the glass half empty or half full?

War Around the Edges

3Com was showing solid leadership as a visionary. Gartner, the technology analyst and advisory firm best known for its "Magic Quadrant" series of reports, positioned 3Com in its higher technology quadrant. (According to Gartner, companies that score strongly on both of the most important attributes for their market—price and performance, for example—are deemed to be residing in the "magic quadrant" for their industry.) 3Com was gaining traction with its boundary routing solution, which reduced the complexity of products à la edge routers. But sales weren't showing the same progress as the technology. And the war with Cisco was beginning to look more like the construction of a large moat around the enterprise customers.

The battle strategy was defined. Cisco had vast numbers of sales ground troops. They were attacking carriers, competing against Juniper Networks, Foundry Networks, Extreme Networks, and 3Com, all far behind them. 3Com was attempting to return to the core from the edge, but Cisco could outman 3Com with its deep direct salesforce. 3Com's sales were strongest in its two-tier distribution channel strategy, which operated more like fulfillment than sales. And Cisco also used a similar distribution strategy to complement its direct sales footprint.

Cisco also had a clever and somewhat predatory marketing practice, leveraging the so-called Cisco Internetworking Operating System (IOS). Barry Eggers, Cisco's first Director of Business Development shared, "They coined the term Internetworking Operating System, IOS. Mostly because people saw Novell had 'Network Operating Systems.' Maybe we could have an Internet operating system. You'd look at our code and say, 'That's not really an operating system.' We called it IOS, we coined the term. It became very popular, so

78. Shore, Joel. "The 3Com Saga: One-time Industry Pillar Hits 25". Network World. April 2004.

people wanted to make sure that they had IOS. They even decided to say IOS was in the switches and all that kind of stuff. All of a sudden, the customer starts thinking, 'Well, I want everything running the same.' It was a great marketing ploy to get the customer feeling like he's got a consistent network even though it wasn't always consistent."

Barry also reinforced their advantage in the enterprise customer space, "The interesting point to make here is that Cisco, by virtue of their router product, and being first to market with routers I would say, became the brand for routers. It was them and Wellfleet, and Wellfleet then had turmoil and Cisco became the brand. So, customers became accustomed to buying Cisco, and if Cisco could deliver something that's "'just as good' in other product categories, they have an advantage with the customer. I think 3Com went to a lot of length to try to build or acquire better products than Cisco, but it was that brand that they had with customers that I think was really hard to overcome." The power of incumbency!

And to cement that tie-in, Cisco also championed a proprietary protocol that addressed the efficient routing of packets. Cisco came out with their own, called Interior Gateway Routing Protocol, or IGRP, and later Enhanced IGRP or EIGRP, which was proprietary, and easier to configure, with better performance, and at the time when others were using less efficient protocols. This helped lock in the customer if they had a Cisco router, to using Cisco devices that would work with IGRP.

"From the start, Cisco was very clever. You could be Cisco-trained as an employee, and they surrounded you as a company. And, you had to think to yourself 'How many John Chambers are there?' John was here yesterday, now he's coming tomorrow to this place, he was everywhere, he must have had a private plane. If we had the tighter relationship with the enterprise customers like Cisco did, due to their footprint in routers, they might have waited for us a bit for multiple products."

– Janice Roberts, 3Com SVP

Chapter 15: What Didn't Happen

Eric had, early on, successfully kept 3Com on par with Cisco in sales and market capitalization as he implemented his vision to streamline the company's products to pursue the global data networking opportunity. 3Com had invested in NetBuilder II, boundary routing, SuperStack hubs, and switches, along with ATM via the acquisition of NiceCom, while maintaining investments in alternative technologies—Token Ring, FDDI, and Fast Ethernet.

And while all the financial figures made it look like everything was fine, earlier actions not taken began to affect the company—most notably, around which technologies and markets to pursue, as well as its product execution.

Specifically, 3Com had decided to not meet Cisco head on in the Fast Ethernet Core Switch arena, and also could not find a way to counter Cisco's direct sales juggernaut, impacting 3Com in its arms race with Cisco. While 3Com had bet on Synernetics in late 1993, an Ethernet LAN FDDI backbone switch, Cisco had bet at the same time on Crescendo. Crescendo had an FDDI hub, but at the time was, more importantly, building a Fast Ethernet switch for the core that used an implementation of Fast Ethernet based on the Grand Junction proposal that had been standardized with IEEE 802.3. Their products were later branded by Cisco in the Catalyst switch family.[79]

While Cisco gets kudos for its direct touch with its customers, their fulfillment of product sales later evolved to utilize big distributors, similar to 3Com's method. The two-tier distribution strategy worked well for Cisco, too, as evidenced by the success of their Desktop Switching Business Unit (DSBU). This unit was created from the Grand Junction acquisition, the company ironically led by 3Com's co-founder Howard Charney along with ex-3Comer Larry Birenbaum and others. This method of fulfillment worked well for the smaller, non-Fortune-100 customers.

In the bigger scheme of things, 3Com's products fit well in the small and medium-sized business markets better than massive complex enterprises. 3Com's product focus led in smaller stackable products in various flavors, such as FDDI, ATM, Token Ring, and Fast Ethernet, but Cisco's large, Fast Ethernet Catalyst Switch went on to be the dominant enterprise product as the Internet Protocol took over the market and became the industry standard for large enterprises.

3Com's answer, with the CoreBuilder 3500 switch, came two years later in the fall of 1997[80] followed by the CoreBuilder 9000 in 1998, using technology from Chipcom and other NSOps divisions, a huge effort to build an

---

79. Molloy, Maureen. "Cisco Buys Crescendo, Plunges into Work Group Switching Market". Network World. September 1993.
80. Daniels, Jodi. "3Com rolls out router replacement". Network World. September 1997.

enterprise-class backbone switch, running at 10/100/1000 Mbps Ethernet speeds. A lead of another two-plus years given to Cisco, in the enterprise space, would spell all kinds of trouble for 3Com within a few years.

One top executive reminisced, "We said to ourselves, 'Man, we're losing to Fast Ethernet. It's not FDDI, it's not ATM.' That we may well have realized too late. We didn't have a central product management group. We only had product management groups in each of the divisions, so there was nobody looking at the entire market, to make decisions about which direction the market was going, and Finocchio is great at many things, but he's not a technologist." Cisco also was trying to keep pans on the fire, buying Lightstream for $120 million for its ATM switching capability in 1995.

Another exec said about this period during 3Com's foray into FDDI, ATM, and not getting on the Fast Ethernet business quickly:

> We had 3Com reps and managers just leaving and joining Cisco with instant promotions. These weren't disloyal employees, but the 3Com chassis that was delivered, did not deliver on the individual modules to go into the chassis. And this was evidenced by the frustration by our 3Com sales teams. Sales guys were leaving due to poor product execution in the divisions. Cisco even had a recruiter in our cafeteria at one point. Cisco would give them live arms and ammunition; their products worked. 3Com was talking about armaments for the future. Our sales execs were not being successful at 3Com but went on to be successful at Cisco, Foundry, and so on. Why? Timely products to market that work.

Matthew Kapp, a leader in 3Com's International Sales during these years, noted that the company needed to put Cisco into a dilemma around their own success, and that 3Com focused on technology while Cisco focused on business and markets. He saw solid success in Asia Pacific and UK in the 1990s, with Cisco copying some of 3Com's playbook, but "the one thing that may have made more difference would have been the appointment of an industry savvy COO in 1995 so we could address market segments with solutions in a much more focused and strategic way, and that aimed to disable, stem, or operate outside a Cisco response."

In a similar but different tone, Ron Sege recalled 3Com's M&A process intentionally lacked an M&A "Czar" or program manager, someone to see over things globally on each deal. "We were intentionally federalist. We had no strong federal government, and all powers were not reserved by the center but delegated to the state."

The Power of Customer Intimacy

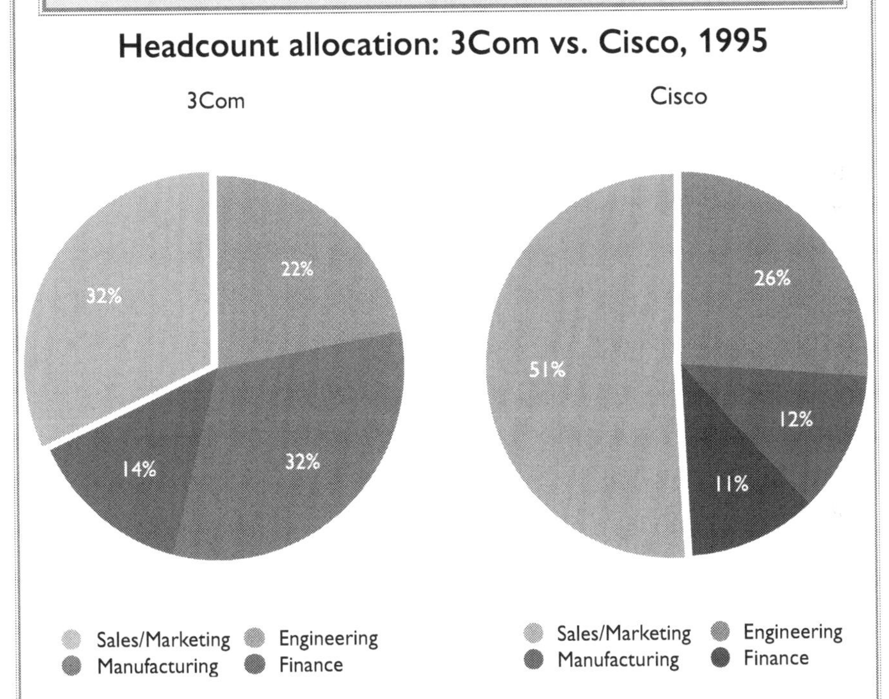

Cisco placed a much higher priority on sales and marketing, with more than half of its people in sales, marketing, and service roles, compared with just over 30 percent for 3Com. This gave them greater intimacy with and connection to their customers. 3Com's headcount was manufacturing heavy. Whatever edge this may have yielded it couldn't counter the effect of Cisco's selling machine.

Based on Cisco's headcount mix in 1995, the difference in the two companies' priorities were clear. More than half of Cisco's people were in sales, marketing, or service roles, compared with just over 30% for 3Com. 3Com, on the other hand, had a much higher percentage of its staff in manufacturing due to having its adapter and lower-end networking products manufactured in-house—over 30 percent, compared with about 12 percent for Cisco.

In 1996, 3Com launched a wide range of initiatives—Transcend Networking (software), a stackable boundary switch called LinkSwitch, under the SuperStack line, and the OfficeConnect products line specific to the small office. These products complemented existing hubs, such as LinkBuilder, NetBuilder II bridge/routers, that connected to NetBuilders running Boundary Routing software, remote access servers called AccessBuilder, and their successful NIC business with 40 percent share of the desktop Ethernet market.

By then, it was evident that Netscape, Mosaic, AOL, and the emerging Internet were beginning to drive demand for networking for the masses. While maintaining the fight in large enterprise accounts, the company was being pulled towards carriers on one side, and consumers on the other. In particular, Eric was feeling pressure from two specific competitors—Cisco and Intel—and felt that 3Com needed more critical mass. Cisco's Catalyst product line was dominating the Fast Ethernet switching market.

The company had come close to buying Cascade Communications, an East Coast company that focused on carrier equipment, but the board and execs eventually backed out. Another potential acquisition—Ascend—was also not to be.

Cascading Between a Rock and a Hard Place

Networking was maturing into the robust industry we know to this day. 3Com's decisions about which companies to buy and what technologies to bet on were pivotal to 3Com's success or failure, but so was 3Com's selling DNA—their two-tier distribution channel, which was low cost and efficient, had been very successful, but was losing ground to the Cisco direct sales train that trumped all. Meanwhile Cisco was cultivating their clout with CIOs, and acquiring great talent.

In 1997, IBM, Bay Networks, and 3Com founded the Networking Interoperability Alliance (NIA) in an attempt to unite as allies against Cisco. The company hoped that by decoupling hardware from software it could grow at the expense of vertically integrated products from Cisco. It did this by testing products for interoperability, allowing for assembly of an efficient multivendor network, as a counter to Cisco's IOS and hardware. Many companies participated, and it helped garner some success. But Cisco countered in 1997 with

"Cisco-Powered Networks," that boosted their brand and strengthened ties with their customers.

To be sure, 3Com had other short-term successes, notably the acquisition of Palm (via the US Robotics acquisition) that generated substantial profits for several years, not to mention the cash generated by Palm's IPO in 2000. But as the traditional lines between voice communications and data communications began to blur, the company found itself with an urgent need to buy a path into the carrier space. And the company never found a successful entry into that crucial market.

As a side note, 3Com did buy an ATM company named Onstream Networks, with estimated revenue of $10-15 million, for $245 million in 1996 to counter Cisco's $4 billion StrataCom purchase targeting the phone companies and Internet Service Providers (ISPs). StrataCom offered both ATM and frame relay switching equipment.[81] But Cisco already owned the space, from LAN hubs to core carrier switches.

During this time, Eric was indeed getting strong pressure for profits and growth from the board. This led him to focus first on Cascade Communications, which was hotter than a firecracker in the fall of 1996. He recalls the reasoning behind their interest in Cascade:

> Shortly before our conversations with each of them, Cisco had purchased StrataCom. This milestone in the industry showed that both Cisco and 3Com understood that networking was going to completely change the world of carriers. Carriers up until then were basically switching circuits, that just carried voice conversations. They were now getting into a data business, and the data portion of the carrier business was starting to look a lot like the enterprise network business. For data networking companies like 3Com and Cisco, it was normal for us to look at carriers as the next big market, but we didn't know anything about selling into that market.
>
> For us, it was clearly the next frontier. I was nervous about this, because it's a very different ecosystem. Carriers had very stringent product requirements, very bureaucratic and lengthy sales cycles. After Cisco bought StrataCom, we inevitably looked at many other players in contention.

81. Internet Pioneer and early 3Com investor Paul Baran helped provide some of the early spark of invention of StrataCom's early products.

So, Jeff Thermond, John Hart, and Steve Russell jumped on a plane and arrived at one of the largest law firms in Boston to sort out the potential Cascade acquisition. Eric wasn't present for the negotiations, having left for China, and was not even reachable by phone. Jeff remembered that meeting, "Their CFO stands up and shows us their quarterly bookings. It's 78 days into the quarter [meaning 86 percent of the quarter had passed], and they are only at 22 percent of their revenue target."

3Com CFO Chris Paisley pulled all the 3Com people aside and said, "First of all, I want to tell you that you've just seen the mother of all insider trading risks and I will personally testify at your criminal trials against you if any of you trade." Jeff Thermond also learned, around this time, that Cascade's CEO Dan Smith was having mammoth doubts about the acquisition. Both Jeff and Chris also knew that Cascade and Ascend had benefited prematurely from RBOCs (Regional Bell Operating Companies, created by the AT&T break up in 1984) buying frame relay switches and other gear ahead of their requirements, and that they were not going to buy more for some time. 3Com would be buying Cascade at the worst possible moment, right before a disastrous quarterly release.

Chris Paisley explained that while Eric was very much in favor of the deal, he waited for Chris to return from a business trip to share his view of the acquisition with the board, showing that there was careful vetting and not necessarily unanimity, and those views should be heard.

In the end, two potential mega-mergers—with Ascend and with Cascade—never materialized. Chris' lobbying efforts based on the numbers were a big part of what quashed the Cascade deal, but there was more afoot. Dan Smith's price expectations were set, but his business ultimately failed to live up to the projections that had determined that price. Eric's disappearance for three days during negotiations also didn't help, and it gave Dan space to question the deal. Cascade began to realize there wasn't anybody in 3Com that understood the business of selling to carriers. Since Cascade was a purely carrier-focused business, their team would also have jumped ship in droves. And buying Cascade just as their quarter was about to tank would have created terrible optics for this potentially monumental technology acquisition. (Cascade ultimately sold itself to Ascend for $3.7 billion in the same month 3Com purchased US Robotics.) 3Com's price for Cascade had started higher according to Eric, perhaps $5-$10 billion. This would have also created dilution on 3Com's earnings and stock price.

Steve Foster, who served as Director of Business Development from 1995 to 1998, noted that in addition to individual acquisition deals of Cascade or Ascend, a three-way consolidation deal had also been contemplated. "That would have made perfect sense. We would have had everything Cisco had, plus more. And Ascend and Cascade got along with each other. But I don't think either of the CEOs were willing to relegate their authority to Eric. They were all relatively young guys, in their early 40s. They were all CEOs of public companies. They had all made phenomenal amounts of money. Strategically, it would have made perfect sense, and we could have struck a deal that worked for everybody—and for the market. But as I've gotten older, I understand completely why the other CEOs ultimately didn't consummate that deal. They were having too much fun and making too much money, and still had other options they could pursue."

Who Said Anything About a Whorehouse?

Although the mergers that many expected didn't come to pass, a sense of change was in the air. Some executives felt that 3Com's cohesive corporate culture was being neglected by senior management, and was starting to unravel. Some felt 3Com leaders had stopped talking candidly about engineering processes and product management inside 3Com. One person interviewed said, "It would be like bringing up a visit to a whorehouse in a Sunday mass—nobody wants to talk about it. There was complaining about why products are late, but no one would talk about the engineering processes. At management meetings like 3Com At The Half, there was good communication between the OpCom and ExecCom leadership teams. It was extraordinarily useful, really well done. I always thought that was really cool, and then it just kind of became this empty shell."

It is true that Eric was not one to hang out with the team; he was always a little distant. When the company was smaller, there was clearly more access to him. And as 3Com got bigger, he was allocating more of his work externally, working with investors and other outside stakeholders. He no doubt felt the responsibility weighing on his shoulders, and he had to make a lot of decisions on his own. Eric would be more isolated as time went on, especially post USR acquisition as Bob Finocchio and other execs departed. At the same time, the company was growing, as was the executive team's responsibilities.

Earlier in history, at the start of his tenure as CEO—with everyone focused on crossing the desert and implementing his global data networking vision—the senior team was smaller, synergistic, and would do whatever it took to cross that desert. In addition, Eric depended more on consensus decision-making in his earlier days due to his lack of experience.

Many believe that changes in the company's decision-making occurred when USR came into the picture. Eric had a very intact team for a good seven and a half years. After the USR deal, the remaining two and a half years were arguably when the executive train went off the rails and the cohesive ethos shifted. And later, after he brought on Bruce Claflin as COO, there was further erosion of cohesion among the group.

Soon enough, emotions among the senior team were running at a consistently high level, and a pervasive cloud of disappointment began to hang over the company. Externally, there weren't obvious signs as to why 3Com wasn't dominating the industry the way that Cisco had done after 1994. After all, the company was led by a team that included many extremely talented leaders. To be sure, there had been mistakes. The Bridge acquisition could have been better managed and leveraged. And the jump Cisco gained with the Catalyst Switch, combined with their amazingly awesome direct selling approach and mind share capture of the CIO and their loyalty was tough to match.

And sometimes there was too much pressure within 3Com to drive consistent earnings across all divisions every quarter. One Wall Street executive identified 3Com's focus on their price earnings multiple—what an investor was willing to pay for a share of stock. In this exec's view, 3Com overvalued the importance of trying to match or compete with Cisco's multiple, and the company was a bit obsessed with gaining Wall Street's mindshare. A fine line exists between having conviction in a strategy versus overselling that strategy. At times, 3Com may have promised things that never quite came to fruition, leading to a lack of trust by consumers and investors.

The tight relationship between the executive team and the second tier of company leaders was becoming brittle. It was devolving into little more than a repeated demand to "Just tell me what the numbers are, that's all I want to hear." Some also described what they called a zero-sum, high-friction game for control between executives like Doug Spreng and Bob Finocchio.

Regarding Doug and Bob, one executive stated, "Management spent so much energy on internal fratricide that should've been spent on annihilating the competition. That was extremely regrettable." Doug and Bob had a different take on the competition between them. Doug said, "When the meeting was over, we got back to work, and I would never believe that Bob was any less committed or productive because of these differences. As for me, ditto. We were company men, first and foremost." Nevertheless, those below Doug and Finocchio felt the rift. Competition and friction among divisions was hardly unique to 3Com, but it may have started ratcheting up at precisely the wrong time.

Back in the USR

The prospect of any deal with either Cascade and Ascend was off the table, but US Robotics (USR) had approached 3Com about joining forces.

"I personally knew two members of 3Com's board at the time. Neither of them had called me, asked me a single question about US Robotics, and US Robotics was having some significant problems, outside of our division at Palm. We were kind of a little gem in there, but the other divisions were having some significant problems that I think 3Com, with a phone call or two, would've been more aware of than they were."

Donna Dubinsky, Palm Co-Founder

Although USR was a widely known consumer modem brand, the company was a telco carrier play and also had products that would help 3Com address the market for dial-up line modems, with prestigious accounts like AOL, a leading Internet service provider at the time. But USR was hurting from changes in distribution channels, a shrinking modem market, and shifts in technology. At the time of acquisition, the dial-up modem market was ready for a speed upgrade—moving to 56K—but was taking a wait-and-see pause to see whether USR's X2 technology or competing approaches from other companies (like Rockwell and others) would win out. But with much faster (and always on) broadband connections looming on the horizon, the entire dial-up modem business wouldn't be worth much for very long. (As an aside, modems have a longer history than you might think; in the 1920s, newswire services used a precursor to modems, which later evolved as encryption devices used by the Allies in World War II.) As it turned out, the most valuable asset from the USR acquisition was Palm, a trailblazing company that had even more of a roller-coaster ride than 3Com.

Chris Paisley didn't mince words in his telling of the USR story, "Probably for the only time that I can remember in any proposed transaction, the due diligence team came back and said, 'We should not do this deal. There are all kinds of red flags in their business.' I remember thinking to myself, 'When the board hears this, there's no way they're going to go forward with this deal.'" Chris went on to explain that Eric did work hard to explain his rationale for his decision to purchase USR, and fully acknowledged Chris' concerns. Eric also addressed his plans for dealing with some of the inherent issues with the transaction with Chris. While Eric may have welcomed hearing Chris' dissent on the Cascade and Ascend transactions, Eric prevailed with the board's decision to proceed with USR over the objection of 3Com's due diligence staff.

Some executives felt that 3Com was buying a company run by a bunch of crooks, who had stuffed the channel, and were dumping the problem on 3Com, and that 3Com had "lost its mind." They felt they were not properly consulted, did not agree with the decision, and made their plans to leave, a regrettable outcome since many of these executives had deep knowledge and the right experience to help run the merged company. No doubt the notion of a "merger of equals" (when it was really an outright acquisition) contributed greatly to this problem.

Steve Foster also had mixed feelings about the USR acquisition. Part of the backdrop to the USR acquisition was a sense of "filling in the picture"—for both the company and for many members of the team. Without the acquisition, they would be left with relatively few options for future growth. Steve Foster said, "It feels like we sort of solved the local area networking problem, we had our own internal developments around Gigabit Ethernet. We also started a venture fund off the corporate balance sheet, and we invested in Extreme Networks and two other Gigabit Ethernet startups" And, Steve noted, "All three of them went public and made us a bunch of money." Leon Woo added to the equation about Gigabit Ethernet at that time, "3Com forged an OEM relationship with Extreme to get an early lead over Cisco in the Gigabit Ethernet race, while our own internally developed switches were a year or more away from shipping".

USR had looked at HP, Cisco, and Motorola, but settled on 3Com as their target to sell themselves to. Casey Cowell, USR's CEO weighed in:

> Cisco had the backbone and we had the wide-area network meaning the public telephone network. From the endpoints on the telephone network, and whether it be an office or a consumer, you're typically going from that endpoint via modems across the telephone network and coming out the other side into our total control hub product. It handled tens of thousands of calls simultaneously and connected you to places like IBM Global Networks, or America Online or CompuServe, or Prodigy.
>
> 3Com had an analogous position to us in the local network world. And Cisco was looking to move out toward the outer edges of these networks, which they eventually did with Linksys, and we were looking for ways to contain them from moving out. And we felt that scale and combining forces with 3Com to combine wide-area network and local area network connectivity was an appropriate-scaled

alignment of forces that could compete successfully, at least at the edges of the network, with Cisco.

For 3Com, Steve felt that the options were more limited and the desire to continue up the growth curve was more urgent. "Eric still wanted to be bigger. He was, I wouldn't say desperate, but he was very eager to try to consummate a deal. And US Robotics was one of the few players of any consequence that he could consummate a deal with."

But while USR may have seemed like a good fit in several regards—its size and product mix, for example—it was a mismatch in many other ways. Steve recalled after visiting the company that, "This doesn't feel right. They had experienced hyperbolic growth, but there was just something off."

On the one hand, Steve noted, "There were some things I really liked about them. When you visited their factories, they had these banners hanging above the manufacturing floor. I think there were four of them, and they had like a dozen names of competitors on each of them. In addition to adding features and using scale to drive their costs down, they had product managers devising specific strategies to beat these competitors. The only time they stopped the massive board stuffing lines was to celebrate knocking off another competitor. They literally drew a line through the competitor's name—and nearly three quarters of them were already gone. It was such an aggressive culture, and I remember thinking, 'Gosh, I kind of love that.'"

But while the aggressive attitude might have resonated with Steve, he had a very different reaction to the company's financial picture. "We did a bunch of diligence, and it was pretty obvious to the accountant in me that those numbers didn't make any sense. It seemed like they were shipping 100 percent of a quarter's sales in the last couple of weeks, which would have been physically impossible to do. We were trying to determine if they were stuffing the channel and taking it back early in the next quarter, or driving modems across the street and calling it revenue, or what."

Steve said, "There were anomalies like that that we called out to management and the board. I really didn't want us to do this deal... And Eric was unhappy because he really wanted to acquire USR—he knew that the Ethernet adapter business was eventually going to a chip on the motherboard, and he needed to have a wide area networking business." As for USR, Steve felt that the USR leadership recognized their core business faced a similar challenge—the modem was eventually going to the motherboard. "They were happy to cash out now and let someone else deal with the challenge."

The 3Com acquisition team expressed their views about USR to ExecCom and the board. Steve recalled, "We presented to the board, and I said, 'Look, I don't think we should be paying this much for this company.' I think what they are doing is not sustainable and we could probably wait, let this blow up and buy it a lot cheaper in a couple of quarters. Eric basically said, 'I don't want to buy a broken company and all of this is fixable.'"

THE WINTER YEARS
(1997–2009)

The Growth Spurt That Sputtered

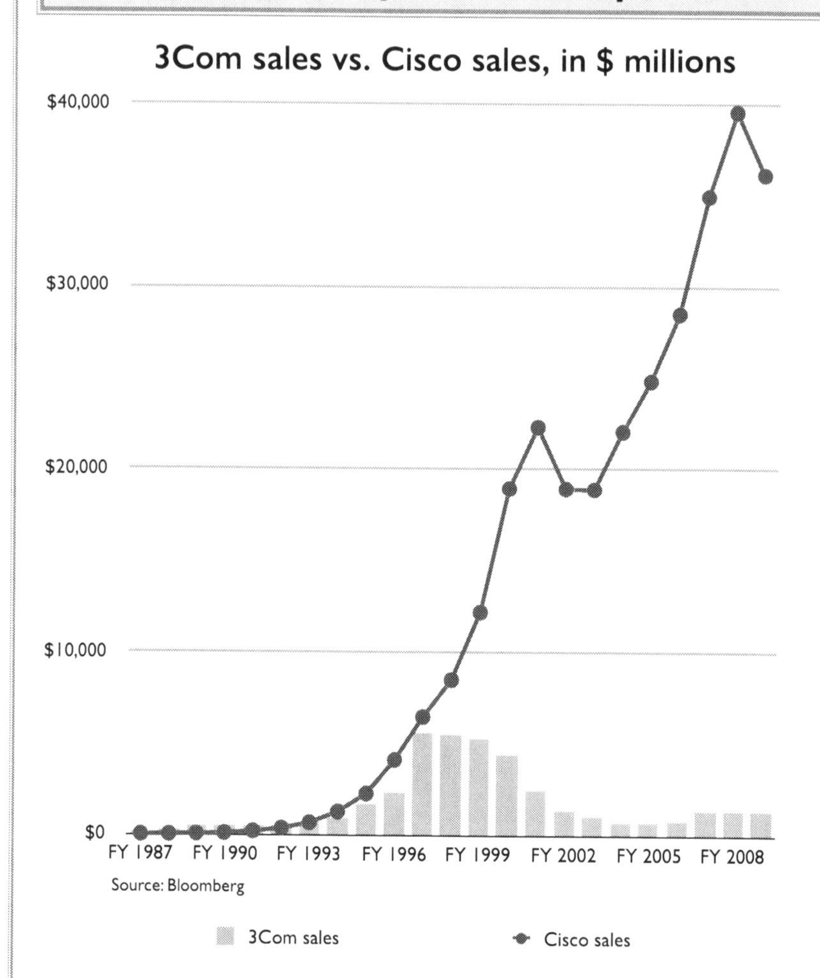

There's an old joke that asks: "How do you wind up with a million dollars in the stock market? Start with two million dollars." That's what happened with 3Com's acquisition of USR. It boosted the company's sales overnight from around $2 billion to over $5 billion—and it only cost them $7.3 billion to make that happen. Unfortunately, sales reached their apex and began sliding almost immediately after the USR purchase. Meanwhile, Cisco was riding the wave of demand for enterprise products and snatching up companies with strong products and technologies focused on growth markets, using its stock price as currency.

A summary of 3Com's last decade: After two decades on a more-or-less upwards trajectory (sometimes more, sometimes less), 3Com went into a tailspin in its final decade. As Bruce Claflin was taking on the role of COO in 1998 (later to become CEO in 2001), the company's ship had moved into rocky waters that started with the acquisition of USR in 1997. In hindsight, the company had few options for keeping pace in an increasingly competitive market that it could no longer easily dominate. Ignoring advice from his team and attempting to move into carrier and telco markets, Eric and the board acquired US Robotics. USR's subsidiary Palm ultimately generated a cool $1 billion in cash for 3Com in its IPO spinoff. And the company also rang up sizable profits when it cashed in on its investments in several startups, including Juniper, Extreme, and Epigram. But, the original rationale for acquiring USR—to double down on its edge strategy with USR's modems, and also fast-track 3Com's entry into the carrier space—fell far short of expectations. Within a couple of years, its product lines were sold off or shut down.

The board and Eric's team also made the fateful move in 2000 to exit the market for higher-end switches, routers, and enterprise products such as CoreBuilder. This move greatly benefited Cisco, as 3Com at that point was the company's last significant, standalone networking competitor. It's exit from the enterprise market was an early Christmas present to Cisco.

To add insult to injury, 3Com along with the entire tech sector were hit hard by an overall downturn in the economy in 2001, adding to the sense of panic and prompting many companies to jettison businesses left and right. Despite this rocky phase, Claflin and the company attempted to reinvent itself again in 2003, with a focus on enterprise markets led by a joint venture with Huawei called Huawei 3Com, or H3C, what ultimately became the "China Out" strategy. After a successful buyout that left 3Com with 100 percent of the H3C venture, Edgar Masri's attempt in 2008 to take the company private (with Bain Capital's help) failed due to the politics of international trade. The controversial appearance of a Chinese company investing in a U.S. company was magnified and condemned by Congress and the Bush Administration. Ron Sege ultimately found a viable exit strategy that led to 3Com becoming absorbed into HP for $2.7 billion.

Chapter 16: USR Aftermath and Aftershocks

Ultimately, the USR deal and its aftermath presented 3Com with all the challenges that the due diligence team worried about, and more. "I'd put my cards on the table," Steve said, "but we went ahead and bought them. So we had to make this work." Steve became part of a team including Janice Roberts and USR's John

McCartney, which was tasked with formulating an integration plan for the two companies. He elaborated, "We identified redundant things—some of the same products. Our recommendation was aggressive—close two factories, eliminate four product groups, and cut 700 or 800 people. Eric said, 'I've told the street that there was not a lot of overlap.' So he decided not to execute that integration plan to its full extent."

"If you look at AT&T and NCR, or IBM and ROLM, the acquirer did not understand that it was acquiring people and a culture. If you don't have a culture that quickly embraces the new acquisition, if you are not careful in the selection process, then the odds are high that your acquisition will fail. You have to avoid the temptation to say, 'Well, our cultures are different, but I can still make it work.' They normally don't."[82]

John Chambers, former Cisco CEO

Steve felt that not acting decisively on the redundancies was the wrong decision. After a large acquisition, he argued, you really have only one opportunity to take these kinds of actions without the fear of negative market repercussions. "You get to say, 'Hey, here's all the redundant stuff.' You dump it in a quarter and then you move forward. It would have been one clean cut, getting clear of all that instead of a bunch of smaller cuts. And you would have sent a strong message to the combined company."

Compounding the challenge of a lingering, slow-motion integration process was a different challenge that was already starting to become apparent throughout the ranks of the company. "Even before we did the USR merger," Steve remembered, "and certainly after it, all of the senior people at both companies were relatively young, most of them were in their late 30s or early 40s. Most people who reported to them were two to five years younger, so if the company didn't keep growing, you were just waiting for that person to leave or die or move to another job. Where people had seen pretty hyperbolic rises in their careers over the preceding five to seven years, suddenly it felt like they were settling in for the siege. I think many talented people at 3Com were feeling like, 'I made my money, now it's a bigger company, and my boss is only five years older than me,' A lot of good people left."

From USR's viewpoint, thousands of employees were let down, due to jobs being eliminated or feeling betrayed by the executives they had been loyal to. According to the *Chicago Tribune* and A.T. Kearney,[83] in a sample of the biggest

---

82. *Inside Cisco.* Page 7.
83. Kidd Stewart, Janet. "3Com-Robotics Merger a Lesson in Frustration". Chicago Tribune. March 2001.

mergers of the '90s, more than two-thirds underperformed their industry peers in stock performance post deal close, and lagged in other measures—sales, profits, and returns. While much of USR's workforce was made up of highly educated, in demand computer professionals, that didn't help the later redundancies for factory workers as manufacturing plants were shuttered or sold.

For many though, at both companies, there were plenty of new opportunities popping up. This was the late 1990s, when the first Internet bubble was expanding but hadn't yet popped. And startups, with venture capitalists to help their recruiting, were demanding 3Com's talent. Foster notes, "I think we bled a lot of talent because the world was different. The opportunities outside 3Com suddenly were exploding. People were looking at their own situations and saying, 'Why not go to Juniper?' or wherever."

Ron Sege also shared his view about the USR acquisition, saying, "Cisco was expanding into the carrier/telco space while we were stuck in the enterprise 'edge in' strategy—adapter cards, wiring closet, SuperStack. We didn't have enough of a carrier class switch. Chipcom, Synernetics and NiceCom helped us with switching, but the idea of USR was to buy it for their nascent DSLAM,"[84] which is basically the line termination of the DSL network on the office side, which USR had just started. We would sell Palm and sell off the modem business, and end up with the carrier product. But DSLAM wasn't proven in the market place. And to get the deal done, USR had stuffed the channel—it was half of what we were told. This created an albatross—shareholder suits, and the like. Eric later went off to deal with Palm and mobile computing."

1998: A Good Year to Get the @#$% Out of Dodge

In January 1998, Eric held an East Coast analyst meeting to review the issues the company was facing at the time, and the strategies to address them. This was a pivotal period. The company had swallowed USR, and the acquisition was not turning out to be all that it was represented. As a direct result of the USR acquisition, 3Com would pay $259 million in November 2000 to settle a shareholder suit related to a period of concealing information in 1997.[85] At the time, this was the second-largest settlement in SEC history.

During this post-USR acquisition period, Chris Paisley and Eric appeared extremely stressed over these issues. Several people in the media had brought

---

84. For more information about Digital-Subscriber-Line-Access-Multiplexer (DLSAM),that aggregates subscriber lines into high capacity uplinks, please visit the Technicalities page on our website at 3comstory.com/tech.

85. In an interesting twist, Abe Darwish VP of Real Estate and Site Services, helped 3Com pay for this lawsuit by selling 50 acres of prime real estate for $220M+ that had been intended for their expansion off North 1st Street in San Jose. A remarkable feat and a great profit for 3Com. The buyer was Palm, although Apple had competed for it as well.

these issues to light, including Herb Greenberg, a financial columnist for the *San Francisco Chronicle*, who had observed that the company's quarterly reporting of inventory-to-revenue figures were out of line.[86]

In fact, USR was hardly in production for the two months prior to the close of the acquisition. Instead, revenues were driven by channel stuffing (filling the distributor channels downstream with excess inventory to inflate current revenues at the expense of future revenue recognition). This was also later documented by *The New York Times*, in their examination of GAAP pooling of interest merger rules,[87] which led to the SEC inquiry and the class action suit.

Some executives mentioned their practice of moving product to a warehouse across the street, then returning product in the next quarter. They also mentioned a practice of "shipping in place" by moving product in their facility "over the yellow tape" from one side of the warehouse to another, for example. If it was on the other side of the tape, then it was revenue. This is a practice no Big Four accounting firm would approve—so they had a local accounting firm taking care of them. Missing a quarter wasn't something Casey Cowell and USR would allow.

Details of the events took over a year to emerge, and almost three years to settle. The merger closed in June 1997, after 3Com's May 1997 quarter ended. While USR had gross sales of $200 million for the two months ended on May 25, 1997, after booking their product returns from the channel stuffing practices plus price protection credits given to customers after their March quarter, sales were only $15 million, (compared to "record" USR sales of $690 million in the quarter ending March 1997) and they suffered a $160 million loss in this two-month period. 3Com had to later restate their results to take the loss through their income statement and lowered their net income by almost $100 million. While not communicated to Wall Street at the time, USR's business was severely impacted by confusion over the 56 Kbps product launch, the lack of standards for that technology, the launch of new X2 technology, and price reductions and other promotions. And USR's significant negative cash flow had forced it to borrow $74 million.

USR execs Casey Cowell and John McCartney (who later received a $6 million severance package in 1998) led the charge in making false or at least misleading claims during securities analyst calls and meetings around this time.

---

86. "Appendix Summary of Cases: 3Com Corporation 97" on corporate whistleblowers. Berkeley Haas School of Business.

87. As an aside, pooling of interests type mergers were very advantageous in the 1990s—they permitted companies to combine balance sheets together without affecting their earnings, versus what was called the purchase method of accounting, which frequently resulted in the addition of a goodwill asset to balance sheets, which could then be written off to earnings. The accounting standards body did away with the pooling method in 2001.

And the analysts bought it. 3Com executives Eric and Chris got sucked into the vortex as they embarked on a road show with the USR team to sell the merger in April and May of 1997, the same period when USR was unloading trucks of returned sales as inventory back onto their docks. Financial firms like Hambrecht & Quist, PaineWebber, and Bear Stearns were all provided with assurances that concealed reality, and those companies published reports that were favorable towards the merger.[88] While 3Com ultimately paid the price twice, (paying first for inflated results, and second for settling SEC lawsuits) it could have been even worse in terms of the company's reputation and the cost of settling the lawsuit.

The top execs at USR also sold substantial chunks of their holdings prior to this disclosure. $200 million in profits were distributed as follows: Casey Cowell ($54 million), John McCartney ($51 million), Michael Seedman ($44 million), Ross Manire ($15 million), and Richard Edson ($4 million). 3Com executives also made stock sales but they paled in comparison to these USR executives' profits.

Merger Mania

In 1997, Cisco's revenues were $5.6 billion and 3Com's were neck-and-neck, with $5.4 billion after the USR merger. But two years later, Cisco had again pulled further ahead, while 3Com revenues began a precipitous decline. And there was a frenetic M&A landscape after the USR acquisition for all networking companies with a real land grab going on. In the eighteen-month period from January 1998 through June 1999:[89]

- 3Com acquired Lanworks, EuPhonics, Smartcode, NBX, and certain assets of ICS and entered into a joint venture with a Taiwanese networking company.
- Lucent Technologies, a telecommunications company, acquired 13 companies, including networking equipment supplier Ascend Communications.
- Cisco Systems acquired 14 companies.
- Nortel Networks, a telecommunications company, acquired five companies, including Bay Networks.
- Alcatel, a telecommunications company, acquired five companies, including Xylan, a networking equipment supplier.
- Siemens A.G., a telecommunications company, announced plans to acquire three networking firms.
- General Electric Company, a UK-based industrial conglomerate, acquired Fore, a networking equipment supplier.

88. "Consolidated Amended Class Action Complaint for Violation of the Federal Securities Laws: Demand for Trial By Jury"
89. 3Com 1999 10-K

- Intel Corporation, a computer components manufacturer, acquired a data networking company and a manufacturer of telecommunications computer components.

Because Cabletron had recently bought Yago Systems, a maker of layer 3 and layer 4 switches to support Gigabit Ethernet, all the major players now had this type of switching. On the one hand, this gave Cabletron a way to flex its muscle as a player against Cisco in the evolving Gigabit Ethernet switching market, and on the other hand, it was an indicator of the critical nature of the arms race with Cisco to shore up weak spots in their product lines.

3Com's product lineup for the moment seemed to be fine, if lacking in home-run hits. But one of the most underestimated aspects of the USR acquisition was the impact of the cultural change to the organization. Eric had earlier said to the team, "You'll love them, their values are the same as ours." His statement had been based on reading the values statement in their policies manual. But after visiting them, many of the executives—Chris Paisley, Eileen Nelson, Doug Spreng, among others—told him that the merger would break 3Com's culture, that USR's culture was antithetical to 3Com's. Eric said that while he respected their opinions, they would be able to manage the cultural differences. The alternative scenario—hunker down and try to slog it out in the marketplace alone, in a smaller scale than 3Com's competitors—would be worse.

The FY 1998 MOST annual operating plan forecasted revenues for the year to be $7.5 billion. Actual revenue for that year came in at $5.4 billion, or 72 percent of the forecast. Net income was $30 million, versus the FY 1998 forecast of $916 million. The writing on the wall, it appeared, was becoming visible.

> **"Culture eats strategy for breakfast."**
>
> *Peter Drucker, Business Management Guru*

The State After the Union

As the 3Com accountants in Santa Clara worked on preparing SEC financial restatements, Eric attended the East Coast Annual Analyst Meeting in January 1998. He had a lot on his mind. He emphasized the following points at the conference:

- The market for converged networks, in which 3Com was the dominant player, was doing well, growing 20 percent a year, up from the 15 percent projected earlier by Data Communications, a market research firm.
- The company was now working its way through a "channel inventory rebalancing issue" (a too polite way of describing USR's channel stuffing). The goal

was to carry only four to six weeks' worth of NICs in the channel, and six to eight weeks' worth of modems. Roughly half a year after the acquisition, the company still had over ten weeks' worth of modem inventory—or almost three months' worth of sales demand—stuffed in the channel).

- In enterprise, the company thought it was doing well—in general terms, hubs were moving to switches, desktops were going from 10 Mbps to 10/100 Fast Ethernet. 100 Mbps FDDI was moving to Gigabit Ethernet and ATM in campus cores. Routed WANS were upgrading to switched WANS. Network intelligence was moving from the core to the edge.
- In carrier systems, which was primarily the remote access products, the company was benefiting from the 56K products, as were Ascend and Cisco. Virtual private networks were growing, as well as telco gear sales for what was called the public switched telephone network (PSTN).
- 3Com was also smartly investing in a Cisco foe called Juniper Networks. Juniper was led by StrataCom founder Scott Kriens, CEO, Marcel Gani, the former VP of Grand Junction, and others from Bay Networks and Cisco, a stellar team, with a very fast router to help eliminate bottlenecks for the carriers and Internet service providers.
- In client access products (modems, NICs, and PC cards), modems would drive higher average selling prices as 56 Kbps products shipped and the ITU standard for 56K would emerge (or so they had hoped).
- PC OEM sales remained solid, with Acer, Dell, Gateway 2000, HP, Hitachi, IBM, NEC, and Packard Bell as customers. The NIC strategy was to "beat Intel," retain the PC OEM business, differentiate with new 10/100 product roll outs, and DynamicAccess software. In FY Q1 1998, they shipped 1.5 million PC cards.[90]

Regarding the modem market, John Hart, who had viewed the merger favorably, noted that the delay in standards approval destroyed much of the modem hype. From an *EE Times* article:

> Taking any kind of decision about 56K PCM modems was difficult at the beginning of 1997 when major modem manufacturers such as Lucent, Rockwell, and US Robotics started marketing two different 56K modem technologies. For almost a year, there was a great deal of uncertainty about 56K modems due to the incompatibility of these two technologies and the lack of a standard. Finally, in February 1998,

90. Michael Dell cited just three companies—Microsoft, Intel, and 3Com—as key suppliers for his namesake company.

> the draft of the ITU V.90 standard was released and it resolved the compatibility issues. The final version was ratified in late September 1998.[91]

John felt their diligence had overlooked the long lead time for this approval, and during that time, the push toward cable modems to the home was on. John recalls:

> I remember an off-site, where we said we better get going on WiFi, the home, but Claflin [as COO and President, given the reins over 3Com's worldwide operations and sales and marketing in late 1998] had no interest. We should have been all over cable modems and saying, "You're in violation of boundary routing, etc." [boundary routers behaved and worked similarly.] We could've locked out everybody else or certainly made them pay a fee.
>
> And so, we lost the home part of it, and shame on us, because we should've owned that. We had won the small and medium-sized evolution, and that got us up to $6 billion, but we missed the next wave. And what kind of router do I have in my house? A Cisco router for my WiFi. That should of been a 3Com. From a networking point of view, we were the fast, cheap, and simple company.

Eric Benhamou had surmised that, in 1999, "USR gave us a way to build up cable modem and DSL expertise. We were increasingly delivering connectivity to small businesses and consumers."[92] But later that path was not the one taken under Bruce Claflin's COO and later, CEO tenure.

What about Bob (Finocchio)?

Another kind of competition became a driving force that shaped the company leadership, and that was internal competition among the 3Com senior team. While internal competition is a factor at play in any large company, it is especially strong between executives brought together through the M&A process. One such rivalry emerged between Bob Finocchio and John McCartney, the USR executive who was tapped as the President of the Client Access unit under Eric. Bob Finocchio was a direct and authoritarian manager—and an enigma to many people. Bob Finocchio was smart, had a strong personality,

---

91. Gao, Frank. "An Introduction to the V.90 (56K) Modem". EE Times. December 1998.
92. Duffy, Jim. "3Com Captain Remains Calm Despite Stormy Forecasts". Network World. July 1999. Pg. 57.

and was a great mentor to many people. But he also had another side, an explosive suffer-no-fools-gladly persona. As one exec recalled, "Bob sometimes had the expression of, 'Why do I have to deal with these people?' He was intimidating to all of us."

Bob Finocchio ended up leaving the company in dramatic fashion. As legend has it, after the USR acquisition, Eric was leading an offsite dinner at the Chantilly restaurant in Palo Alto. During dinner, Finocchio became incensed around concerns of USR's integrity and principles, as well as a potential culture clash. In Eric's view, Finocchio's comments were negative, judgmental, and not characteristic of an executive leader. Eric expressed his dismay over Finocchio's remarks, left the restaurant, and Finocchio's career at 3Com was complete.

In a darkly humorous twist, since Finocchio had caught a ride with Eric that evening, he had to find his own way home. He hitched a ride with Janice Roberts, who recalled Finocchio saying something along the lines of, "Well, I don't think that went so well." Eileen Nelson tried to persuade Eric the next day, that "Yes, he's provocative, but that is his behavior. He cares deeply about the company." Eric responded, "Eileen, you misunderstand me—you think I don't know that? But his behavior is destructive, and I can't have this continuing."

Here is how Finocchio described the events of the evening:

> Eric was talking about how we should run the business after the acquisition. And I had just flown in from a long business trip, I had a bad cold, I was tired and grumpy. And at that dinner I got into it with McCartney about how it ought to be run. And I wanted control of the systems part of the business, you keep the modems.
>
> And I didn't behave well that evening, and it was clear there was going to be a lot of tension between me and McCartney. We were arguing over the remote access business being part of the systems business I was running. And I was getting a little grumpy with Eric, and I was at the end of my rope with Janice and some of her stuff. Because it was all kind of high-level BS. So, it was a very tension-filled dinner and I got Eric very mad that night because I did not behave well.
>
> So, over the next week Eric and I had a couple discussions, and we came to the conclusion that, "This is the way I feel about these guys, and clearly I'm not going to be able to work with them. And we're going to do the deal, and he's going to be on the management

> team. Maybe it's time we part." So that was in April or May of 1997, before the deal closed. My assessment of the USR people turned out to be correct.

Finocchio left and subsequently became CEO of Informix, where he had the opportunity to lead a company that was struggling, put his own mark on it, and vindicate his standing as an effective leader. Given the shenanigans he faced at Informix, it was no easy company to handle, either. But he remained CEO for the next two years there and then moved into the role of Chairman for another year.

Others shared their thoughts on Eric's hiring and firing habits, noting that sometimes Eric would compromise on tough decisions about people and redundancies in the aftermath of acquisition. People intuition wasn't Eric's strong suit. And it's regrettable that Bob Finocchio left while USR executive John McCartney remained, a member of the channel stuffing executive team that caused financial restatements, lawsuits, and the SEC investigation of 3Com (not to mention a tremendous mess for Eric and Chris Paisley to clean up).

Finocchio handed the Network Systems Group off to Ron Sege, a seasoned 3Com executive that would later return to be 3Com's last COO. Leon Woo, a founder and architect for Synernetics' successful switching products, recalls the difficulties faced by Ron Sege when taking over the Switching Division on the East Coast:

> It really was 3Com's lack of systems focus articulated by some of the 3Com leadership team that spelled doom to the Switching Division in Boxborough (then Marlborough). I recall vividly that Ron Sege as [President] of the [Network Systems Group] gave a speech to the employees circa 1997 where he said, "The only thing that matters in networking is cost."
>
> The reason it was so vivid in my mind is that it generated a parade of engineers into my office asking me if 3Com's Switching Division really mattered at all to 3Com. Our products were the costliest in 3Com. I tried to explain that that was motivation to reduce the cost of our systems products, not to abandon them. But, first, with HR not incentivizing engineers to stay (scant stock options or equity incentives, and no sabbatical years' grandfathering over from their prior startup years), second, the apparent lack of senior management's support around the importance of systems products, and probably most importantly, third, the allure of going to a startup, created the perfect storm for staff leaving 3Com.

Making Sense of it All, or Not

As a way of explaining the differences in the two companies, Tracy Lunquist, a USR employee that worked in Marketing in the mid-1990s, weighed in, "Both companies were, as I saw them at the time, tuned into the wants and needs and style of their target customers. Enterprise expected a suit, and 3Com wore one. Consumers expected a polo shirt and khakis. and USR dressed accordingly." Tracy recapped the USR view:

> 3Com didn't understand how we worked at USR, and didn't particularly care. The new building came with 6-foot high cubicles that isolated us from each other but didn't block noise (the worst of cube life and the worst of office life all blended together). We were told not to put anything on top of our wall cabinets or outside our cube walls. The place was like a rat maze; it could take a considerable while to find anyone in the identical endless aisles. On a business trip that summer, I learned that Santa Clara HQ was even worse. It was relentlessly beige and enormous and lifeless—except when we got our summer thunderstorms and discovered that those gigantic walls of windows leaked like sieves. Then it was exciting![93]

And as with Bridge, the size disparity between the two "merged" companies created unrealistic expectations. As Eddie Reynolds put it, "The deal with USR was also called a merger. In reality, we were taking them over, but we didn't want to say that in the beginning. This led to problems after—teams thought they had power, when they didn't. They weren't always playing by 3Com's rules; they thought they could still be USR. But that wasn't the case at all."

3Com also managed its business with a very high level of integrity and a conservative bent. USR, under their aggressive management team, in contrast, was more inclined to take all kinds of risks. USR took risks in dealing with their channels, in inventory, as well as in pricing.

The sheer size of the transaction was also daunting, although Eric had a "divide and conquer" mindset about it. He felt they could separate the chunks into three; first Palm, on its own track, second, put the modem business with the adapter business (Personal Connectivity Operations), and third, carrier products, into the systems business:

> Every chunk was, by itself, a very, very difficult business to integrate because it required us to excel across a broad range of activities. It was stretching our brand.

---

93. "3Com: Crossing the Bridge". The Magic Soapbox. October 2012.

> So, we simplified the problem by saying, "Okay, Palm will have its own brand." Then we tried to integrate the USR brand and the 3Com brand, under the 3Com brand. We kept the USR modem brand, because it was consumer-oriented, but everything else became 3Com Blue. At the time, 3Com was clocking around $1.5 billion in revenues each quarter—the same run rate as Cisco.
>
> But at the same time, the world was discovering the Internet, and dot-com companies were going public at obscene valuations. The venture industry was going crazy, and all the reasonable acquisition targets had unaffordable valuations. We could not sustain that.
>
> But ultimately, that wasn't what broke the strategy. It was because we were still intent on not only preserving, but expanding our position in the enterprise. Cisco and others were extremely strong there. Most of the product innovation was being driven by major platform upgrades. Our enterprise platform for the core of the network was called CoreBuilder.

While CoreBuilder may have had its merits, the most urgent battleground wasn't just a head-to-head product showdown, it was a showdown between the companies' front lines in sales and service, which Cisco excelled at, along with their jumpstart with entrenched lines drawn around the enterprise CIOs.

Slow Progress in Fast Ethernet

The market was moving to faster flavors of Ethernet, and Cisco seemed to be making all the right bets on what flavors the marketplace wanted. Executives noted that, at the time, 3Com resources were spread thin across numerous technology platforms, and the company was left underinvested in Fast Ethernet. 3Com's federalist nature also led the company to Invest a bit inefficiently—understandably, but regrettably, there were three 3Com divisions working on the company's next generation of products that would feature Gigabit Ethernet. And finally, others felt that 3Com's legacy architecture contributed to the company being trapped by its success in either FDDI and/or the successful smaller stackable products.[94]

Even if 3Com had executed flawlessly on product innovation, the outcome may have been predetermined. As Eric put it, "Despite the fact that Cisco didn't

---

94. If you are interested in the nitty-gritty particulars of the Fast Ethernet horserace, you can find some of the details described by Andy Gottlieb a former Apple engineer turned 3Com Product Manager, and Dave Tolwinski from the Switching Division (Synernetics) at 3Comstory.com/tech.

When Customers Speak, Wall Street Listens

3Com vs. Cisco market capitalization, in $ millions

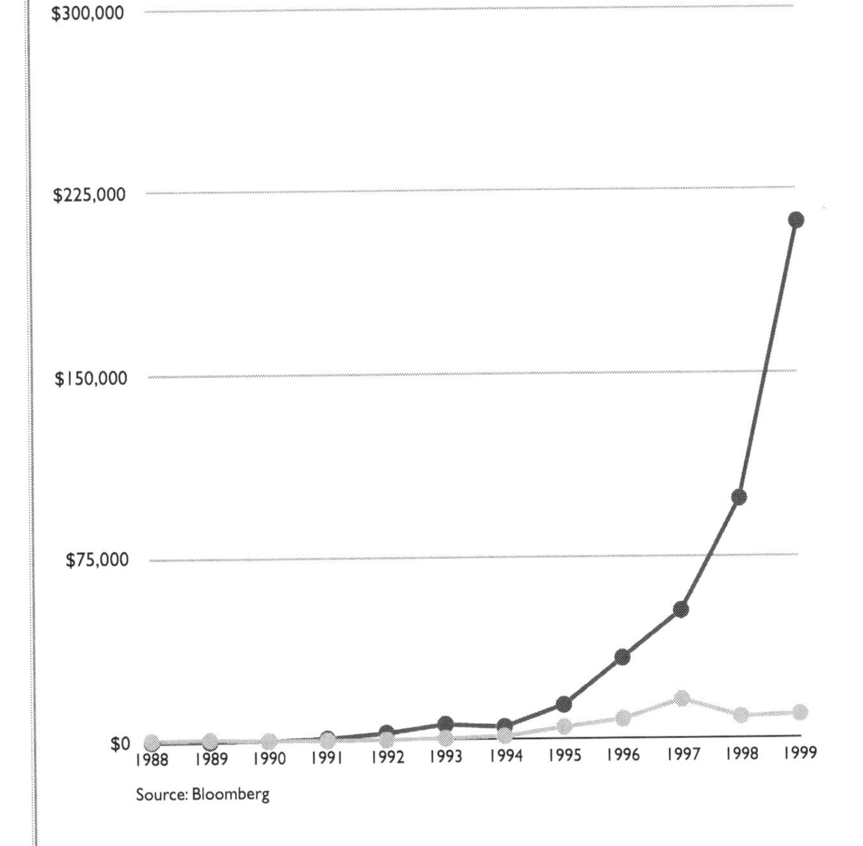

By 1995, Cisco's lead—and momentum—had become unstoppable. Customers' buying decisions were being driven by key products where Cisco had a clear (or perceived) edge and a tight relationship with its customers. 3Com may have expected to at least pick up sales in categories where it had superior products. But once the customer had settled on key Cisco products, they often filled out the rest of their shopping cart for the same vendor, rather than choosing best-of-breed products for all of their different needs.

have a best-in-class product in many categories, their success was, to a large degree, a result of having gained mindshare among the key CIOs of the largest enterprise accounts. Once they had this position, they no longer had to have the best product, because the CIOs always knew that Cisco would take care of them. This was similar to the role that IBM played 15 to 20 years before, in the times of the mainframes. Up against that threat, there was no reason to fund yet another ambitious R&D program." After 1995, Cisco's dominance and market capitalization were unstoppable.

Wall Street had concluded that 3Com's value was quickly fading in comparison to Cisco's. The market capitalization of the two companies shows the disparity.

Comparing the number of employees for the two companies from 1988 to 1999 also provides a sense of the trajectories for the two companies:

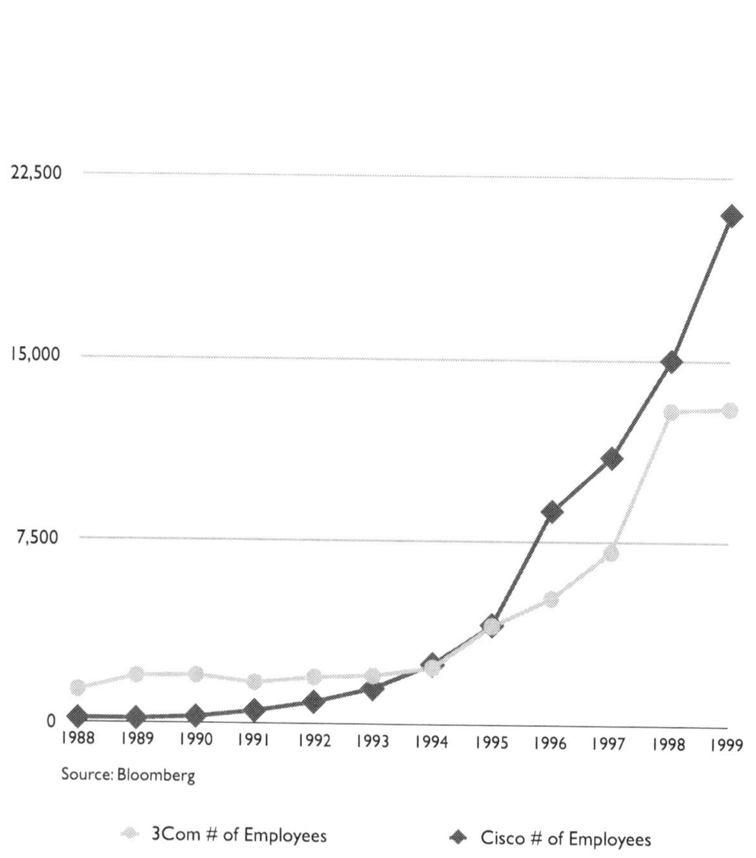

3Com had more employees than Cisco until 1995, when Cisco's rocket really launched on the success of the Catalyst business.

And looking at sales per employee, extended all the way through 2009, it is clear that Cisco was much more efficient at leveraging its headcount to drive higher revenue than 3Com.

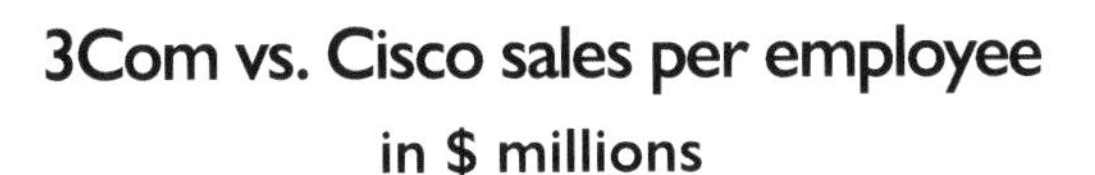

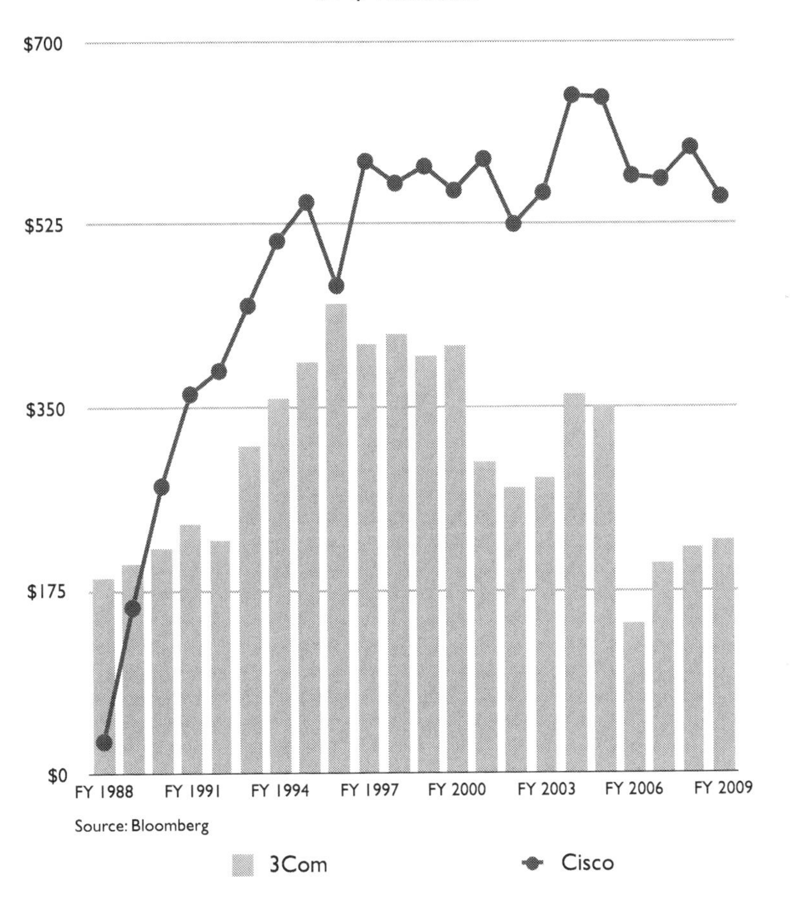

Cisco's productivity outpaced 3Com's; Cisco also used more outside contractors and an outsourced manufacturing model, which 3Com adopted in FY 2004.

Forfeiting the Enterprise Prize

While 3Com led in NICs, desktop and workgroup switching, channel distribution, and small and medium-sized businesses, it was weak in core backbone switches, direct sales and support, and sales into large enterprises. The CoreBuilder was a very ambitious program with new backplane switching technologies, ASICS, power supply—all very high-end.

"We missed, in the past six years, two key cycles: basic high-density chassis with nice routing technology and 10/100 in the core. I believe it's an unforgivable mistake for a company like 3Com. We have corrected our miss with CoreBuilder 9000, one of the better platforms out there for high availability and Gigabit Ethernet switching."[95]

Edgar Masri, VP of Enterprise Systems, in a candid interview with Network World in 1999

But 3Com was coming from behind. The Dell'Oro Group reported that 3Com's larger overall market share of layer 3 switching decreased from 20 percent in 1998 to 6.4 percent in 1999, while its layer 2 switching stayed close to a flat 15 percent. CoreBuilder did not deliver as expected after launch. Ultimately, it had become little more than a very expensive R&D program.[96] As Eric explained:

> At some point, our board became a bit scared about the level of investment involved. I sensed that, despite the confidence I had in our technical ability to come through, the proposition to continue to go toe-to-toe against Cisco in the high-end enterprise became untenable. Not for lack of technical capabilities, but mostly for lack of complete commitment.
>
> Our business had become large, complex, diversified, and perhaps a bit overwhelming for our board. I sensed that they were not going to be supportive of this ambitious approach forever. 3Com was still a very profitable company, with gross margins approaching 60 percent and operating income in the high teens, so a very attractive financial profile, even by today's standard. I almost felt like our strategy and our capabilities outgrew our board. Our board was putting more and more pressure to simplify the business.
>
> Our board meetings were getting longer and longer, and, more often than not, the conversation degenerated into a familiar refrain: "Well, Cisco is clocking 65 to 70 percent gross margin, why can't

95. Network World Editors. "3Com Net Chief Speaks Out". Network World. March 1999. In a candid interview with Network World, Edgar Masri, SVP of 3Com's Enterprise Business made these realistic assessments.
96. Duffy, Jim. "Big Changes Brewing for 3Com" Network World. March 2000.

we do the same? Cisco is doing 30 percent operating income, why are we lumbering along at 18 percent?" Of course, we were very different businesses and we had an altogether different sales strategy. It was hard to pull this company forward and integrate all the pieces.

Finocchio had left the company already, and he was my most senior executive. He was the only one who really understood the whole picture with me. Chris did too, but only from a financial standpoint. Our core systems strategy had really been led by Finocchio. The board being a bit wobbly on us, and with the immense amount of work that I personally had to shoulder, I figured there's no way I can keep doing this. I had to figure out a way to simplify the business.

I had to pull back from the big bet that we were making on CoreBuilder 9000 and the large enterprise business. We'd felt that you could rationalize the expenses for developing the higher end products because those technologies eventually would migrate to SuperStack.

The combination of these things led to a crucial decision of pulling out of the CoreBuilder 9000 program, which was the most expensive program ever undertaken by 3Com, involving probably 10 divisions. Pulling out of it had the effect of not merely abdicating the lion's share of the high-end of the enterprise business to Cisco, but ceding to them the most lucrative part of the business.

This was the most difficult decision I'd made as CEO. It was more difficult than the Renaissance Plan, it was more difficult than crossing the desert, our first acquisition in Europe, the first acquisition in Israel ... all these risks paled in comparison.

Hindsight is 20/20; I wish I had a chance to take this back. I was traveling quite a bit, 80-hour weeks for over a decade. Perhaps if I had a little more mental energy, I would've taken on the board and told them, "Listen: This. Is. The. Endgame. This is not a time to pull back; this is a time to double down. Yes, so maybe we'll have a couple of wobbly quarters, maybe we'll have a surge in operating expenses. But this is where the endgame is played."

> But I didn't do it. As a result, the strategy took a different direction, which was clearly not oriented on the high-end of the enterprise. Instead, we limited ourselves to the mid-enterprise—still a good business, but not nearly as lucrative or as ambitious.

McKinsey and Company, in addition to advising on the Palm spin out was also asked to make strategic recommendations on optimizing value from 3Com's different businesses. In hindsight, their recommendation for 3Com to exit the enterprise business was disastrous, and did not help 3Com get out of its problems. McKinsey suggested that 3Com's layer 3 switching products had fallen behind the competition. Their analysis was basically, "Extreme and Foundry are eating your lunch, so if you can't buy them to upgrade your layer 3 switching product line (because they had very high valuations at the time), then sell the layer 3 switching business and be just a layer 2 switch market to the small/medium businesses." With the clear vision of hindsight, those in favor of the exit now agree that this was an inane and most ill-conceived suggestion.

Why? 3Com lost its ability to serve customers that wanted a total edge and core networking solution. As a result, 3Com's revenues in the overall enterprise business tanked. As Anik Bose, hired later to run Business Development recounted, "A $2 billion business went to $1 billion within 7 quarters, and all the layer 3 switching business went to Cisco; what didn't go to Cisco, went to Extreme. Every order with a layer 3 switch at the core and layer 2 at the edge went out the window. The unintended consequence of the Palm shareholder value-maximization IPO transaction was distracting management's attention from some of the core businesses. At the same time the transition to LOM [LAN on motherboard] being driven by Intel was also cannibalizing 3Com's NIC business. NICs went from $1 billion to a few hundred million in 18 months." The decision ultimately was very damaging.

Bruce Claflin recounted the day when the team decided to exit. He noted Eric had convened the team to discuss and weigh in, and that no one had seriously objected to it. He himself was in favor of the decision, and he complimented Eric for hearing folks out. However, Edgar Masri made it very clear that he had been very much against it.

Out of 21 ExecCom members, Edgar Masri, the VP running the enterprise business at the time, had been the only one against the exit. But Masri was unable to sway Claflin or the others, and he also noted an important feature:

> McKinsey was careful to state, "3Com to exit silently the enterprise business." Some of the key execs at 3Com chose to ignore the word

> silently and believed the imminent Palm IPO and company valuation would give them room to fully restructure the enterprise business.

Masri felt that you couldn't sell work group equipment separately from the data center. And indeed, Masri found that when 3Com got out of enterprise, their market share for the workgroup business dropped from 24 percent to 16 percent. At the same time, Cisco's workgroup business increased from 50 to 56 percent, while Extreme and others gained two percent.

Howard Charney made the following points about Cisco and 3Com's strategic misfire, relating to customer focus:

> Well, what Cisco did was come out with an ever more powerful, ever more increasingly functional set of routers and switches, came out with switches that we called fixed configuration switches like 24 or 48 ports, and then came out with a big chassis where you could put cards in them and the cards were line cards, but the cards could also be cache engines or network engines of some sort, X25, whatever you want. ATM engines. So, Cisco came out with this range of switches and then [importantly] routers that 3Com could have competed with if they would have made the decision to do it earlier.
>
> 3Com lost its strategic way. It lost the appreciation for: What does the customer think is really, really important? What is strategic, not tactical, strategic to the customer? Who is the customer? Oh, it's the largest banks, it's the largest enterprises in the world, say. What are we going to sell them and where do they have angst about their procurements?
>
> Cisco managed to place itself at the center of that critical decision-making process and 3Com wasn't even in the hunt. It was as if it wasn't there. So, when you point and say this year 3Com was X and by that other year Cisco was 5X or 10X, it's because of the strategic nature of routing versus switching.

Tom Peters, the business management guru, wrote something in the foreword to Rich Karlgaard's book, *The Soft Edge*, about HP's changes in its later years that also rings true about 3Com, "To be sure, the HP Way took a wrong turn with a succession of CEOs that managed by the numbers and strangled the essence of HP." In a classic innovator's dilemma, 3Com remained focused on its historically most profitable products, and missed out on potentially game-changing disruptive technologies.

Besides jettisoning CoreBuilder, its enterprise backbone chassis switch, the company also dropped NetBuilder, the essential routers that its customers relied on in the enterprise for WAN connectivity. Just as the Internet's growth was exploding—and driving an urgent need among enterprises to connect to the big-dog central switch—3Com was stepping away from that market.

"3Com's CoreBuilder exit was basically walking away from the strategic nature of your value to the largest customers in the world. Do you know what a heyday that created for Cisco? Cisco account managers could go into customers and say, 'You know those CoreBuilders you have? Well, guess what. They are end of life.' And they go, 'End of life, what?' Even if they're still sitting there working, we will help you get rid of them because they're getting out of that business and it's turning over to this other company and it was such an easy sale. Cisco sold millions and millions and millions of dollars of routers as a result of that decision."

Howard Charney, 3Com Founder, Retired SVP, Cisco

Chapter 17: Simply Palm

Palm, acquired via the 1997 USR merger, ultimately became one of the biggest operations within 3Com for a short period. Because it fell completely outside the networking market, it operated largely independently. It proved to be a lucrative part of the company, but also brought its own set of challenges to the table.

By 1999, Palm's revenues had swelled to $564 million, and nearly doubled the next year (over $1 billion). Tax complications rendered Eric unable to sell or spin out Palm in those first two years, as well as his own internal compass not wanting to spin it out so quickly. Palm executives Jeff Hawkins and Donna Dubinsky, impatient and upset with 3Com's direction for Palm, left to start a new company, Handspring, which led to its own set of challenges.

For 3Com, the revenue and the enhanced perception of 3Com as a market innovator that Palm provided warranted keeping Palm close to the mother ship. Founded by Jeff Hawkins, Palm was incorporated in 1992. As Judy Estrin outlined in her book *Closing the Innovation Gap*, "Hawkins embodies the core values of innovation. He has always been intensely curious,

The Customer Is Always Right, Even When They're Wrong

Switched Ethernet sales, in $ millions

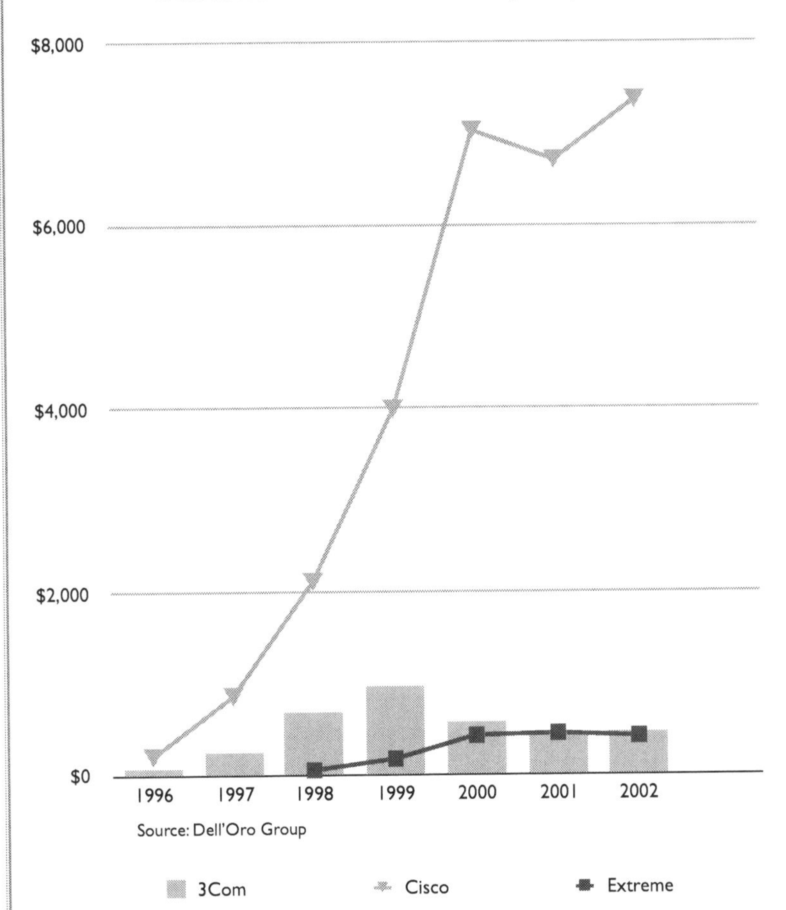

When 3Com exited the market for enterprise products in 2000, it didn't just forfeit the sales for those specific products; it also hurt its sales in its remaining product lines (such as in stackable or workgroup solutions). Customers began relying on Cisco as a one-stop shop (which 3Com was no longer). And even smaller customers who didn't need heavy-duty enterprise products were often swayed by Cisco's growing reputation as the market leader.

he is willing to take risks, he is open to change, and he is tenacious and patient, gaining the trust of others while never losing faith in his own vision." Despite the company's impressive technology, it faced liquidity problems and was unable to attract venture capital after the boom and bust of "pen computing" devices and platforms in the early 1990s. Jeff recruited Donna Dubinsky and Ed Colligan to the firm, and they guided Palm through the launch of the Palm Pilot. The team remained active after they sold the company to USR in 1995. In 1996, Palm sold its first handheld computer, quickly establishing a significant position in the market for what was often called PDAs (personal digital assistants, the precursors of today's ubiquitous smartphones).

In 1999, after waiting the requisite two years to avoid adverse tax consequences, and to further increase its shareholder value, 3Com announced its intent to separate the Palm business from 3Com's business by spinning it out as an independent company. Hawkins and Dubinsky had already left Palm, frustrated that they couldn't persuade Eric to take the company public sooner. They also felt that 3Com's timing was intended to trounce the products that their new company, Handspring, was announcing. While this was very frustrating to the Palm founders, it was definitely profitable for 3Com to hang on to Palm for another year before triggering the IPO.

Janice Roberts added value to Palm by leading marketing programs, such as the provocative "Simply Palm" ad campaign in 1999 that featured naked women holding Palm devices. Not only did this garner attention, but it also generated additional consumer awareness while the press debated the merits of using naked women in advertisements for technology products.[97]

Donna's Dismay

Donna Dubinsky recalled the environment after 3Com acquired USR, for the short period before she left:

> When I was running the Palm division, I often received edicts from 3Com management that seemed crazy to me. For example, when 3Com ran into trouble, it started cutting everybody's budget. I'd go to them and I'd say, "No, no, no. That's not very strategic. You want to cut the places that are losing and you want to feed the places that are thriving and that are winning. You don't cut back the places where you've got huge success and huge opportunity ahead of you.

97. For more insight into the Palm and Handspring story, check out Piloting Palm, written in 2002, by Andrea Butter (a former Palm marketing executive) and David Pogue (the popular tech journalist), about Palm's earlier days.

> That's the future." But they said, "You're not a team player, so you need to cut 10 percent like everybody else is doing." I did not feel that this was a strategic allocation of resources.
>
> If Eric hadn't put us under Janice, things would have been far, far worse. Janice did a very nice job of trying, and succeeding to a certain extent, to protect us and saying, "Look, these guys are little. They need to be nurtured. They need to be grown. They can't be crushed by the whole of 3Com here. Don't go in messing with them."
>
> We ended up leaving because we felt we couldn't succeed. That was why we left. And the main reason why our bankers recommended letting us spin off earlier, is that they said to 3Com execs, "You're going to lose them, and then you're going to fall behind. These people are so critical that you can't do that." But 3Com management didn't accept that. They thought anybody could do the job, which of course, anybody couldn't, and Palm ended up really getting into trouble.

Donna shared her take on an interesting strategic blunder by 3Com in separating hardware from software:

> One of my biggest strategic disagreements with 3Com management while I was at Palm, was they wanted us to license the OS, and they said that we were too Apple-thinking, and that we needed to be more like Microsoft, and be the OS of this next generation of devices. License the OS to all comers. We didn't think that was right for a variety of reasons, and we argued about it with them constantly. Our views had to do with how highly integrated these devices were, that the OS and the hardware needed to be highly integrated, and that the industry was too young, and that it required holistic innovation, and so on. This was a strategic disagreement. So then when we left we said, "If you're keen on licensing the OS, why don't you start with us? We'll take a license." And so they agreed. At that point, it was hard to argue against licensing after they had been arguing with me about this for so long.
>
> So, we negotiated a license with our former colleagues, and left to form a new company, Handspring. After we shipped our first devices, we took 30 percent of the market away from them, literally within

> a year of shipping, and that was when they decided it wasn't such a good idea to license the OS and they decided to stop licensing the OS!

The Spin on the Spinoff

The decision to spin out Palm was fatefully decided by a push from the board. Mark Michael, corporate counsel, recalled some of the drama of that day:

> The key to the chain of events was tied to the board review of the annual operating plan. The proposed operating plan [FY 2000] was disappointing to the board to say the least, which is why the lead outside director [former Wells Fargo president] Bill Zuendt summoned me to the door of the board room a half hour before the scheduled start of the board meeting, and asked for a clear explanation of his powers, duties and responsibilities. Shortly thereafter I was summoned again and requested to call Larry Sonsini and arrange for the board meeting to continue at WSGR [Wilson Sonsini Goodrich & Rosati, one of the premier legal firms in Silicon Valley] facilities. The meeting reconvened at WSGR very early the following morning at 6:30 a.m. with the topic being board authorization to initiate the Palm spinoff.

This story is not without controversy, and it is not meant to make Eric appear reluctant in selling Palm. Rather, according to Mark, such a sale was not initially in the plan, and so the board took the unusual action to force the decision. Eric confirmed that both he and the board had concluded, in the preceding year, that Palm was developing into a business that had little connection with networking infrastructure and was potentially diluting the 3Com brand. With the board's "real-time" feedback and Eric's full agreement, the company decided the time was right to spin out Palm.

Eric asked Goldman Sachs to evaluate a spinout of Palm, which subsequently went public on March 1, 2000. The timing seemed perfect—the NASDAQ was rallying around the 5,000 mark—and everyone was feeling pretty good about where they stood. Unfortunately for Palm, the newly independent company was about to lose its footing.

Palm Loses Its Grip

After Palm's IPO, 3Com distributed its remaining shares of Palm common stock to 3Com shareholders in July 2000, handing out roughly 1.5 shares for each

"After we [the founders] left, Palm essentially did minor, incremental improvements on the product, while the competitors were doing substantial new functionality. A company like this doesn't fail overnight. It just starts to lose steam when it is no longer competitive based on several years of lackluster product strategy."

Donna Dubinsky

share of 3Com they owned. At the time, this was a lucrative transaction for shareholders, but Palm crashed back to earth pretty quickly. After trading at $95 on its first day of trading, it closed at $6.50 by June 2001, 16 months after their IPO, after encountering all sorts of issues. Based on the $95 stock price, the market had valued Palm at $53.3 billion, which was ludicrous relative to 3Com's valuation at that time of $28 billion. Since 3Com held 94 percent of Palm's shares, it should have been valued north of $50 billion.[98] Although this pricing anomaly later self-corrected, it also irreversibly tarnished the reputations of both 3Com and Palm.

Does Eric deserve large credit for Palm's financial payoff, that put $1 billion plus into 3Com's coffers at the IPO? Palm sales had expanded from $563 million in FY 1999 to $1.057 billion in FY 2000, or an 87 percent increase. All of this contributed to Palm's enormous valuation at IPO. But in retrospect, as Palm perilously split the hardware from the software, founders Dubinsky and Hawkins angrily departed, and new CEOs turned over, one has to wonder if it was just used as a short-term cash cow milked to help the rest of 3Com offset its decline in other businesses. What if Palm had been spun off earlier with Donna Dubinsky and Jeff Hawkins at the helm, and they had led the charge towards devices that would have preceded the iPhone?

After the founder left, Eric hired Robin Abrams, an Apple executive, to run Palm but late one evening, five months later, Robin called Eileen Nelson and left a message that she was resigning and going to another startup. Eric asked Alan Kessler to run Palm as an interim President. Alan had already been heading up the Palm Platform group. Alan Kessler was liked by many and brought a steady and experienced hand to the role.

---

98. Norris, Floyd and Lawrence M. Fisher. "Offspring Outweighs Parent as Offering Hits the Market". New York Times. March 2000.

Carl Yankowski, brought into Palm in December of 1999, had been an executive responsible for consumer products at companies like Polaroid, Sony, and Reebok. Executives indicated that very few members of the board at the time actually interviewed Carl. If they had, perhaps they would have decided to not hire him. From all accounts—including interviews with Palm executives—he was cut from a different cloth than typical Silicon Valley executives. That's both a metaphorical and a literal evaluation—you can read about Carl's famous gold pinstripe suits in Andrea Butter and David Pogue's book, Piloting Palm, or on the Wikipedia entry about him. He was smart (in an old-school Boston way), well-connected (and connections meant a lot to him), and moneyed (and showed his money). Culturally, he did not fit in well in Silicon Valley.

With the spinout from 3Com, Palm was just getting started with a string of corporate metamorphoses. In December 2001, Palm formed PalmSource as a stand-alone subsidiary run by Alan Kessler, and PalmSource was later spun off in October 2003 to develop and license the Palm OS. Todd Bradley was brought to run PalmOne, the hardware side of Palm. (Todd left Palm in 2005 when HP's Mark Hurd hired him to become EVP of HP's Personal Systems Group, which then purchased Palm five years later.)

Donna Dubinsky recalled this decision to split the hardware from software as "the ultimate downfall of the business." She continued, "The main reason why Palm is not where Apple is today is because of this poor strategic decision." Point well taken. In the PDA and mobile device market, keeping a vertically controlled system together works better. Apple's vertically-organized iPhone, iPads, and Sony's PlayStation are consumer market examples where keeping the parts together works best.

After a circuitous route of a 3Com divorce, a Handspring remarriage, trying to make ends meet, and all the usual trappings of a challenging relationship, Palm was ultimately bought by HP for $1.2 billion in July 2010. Leo Apotheker, HP's CEO at the time, indicated HP felt Palm's built-for-mobile webOS operating system was the key asset and motivation. But he didn't remain HP CEO very long—partly as a result of this decision. Ultimately, on Meg Whitman's watch, HP sold the webOS to LG Electronics in 2013. HP sold the Palm brand to TCL/Alcatel in 2014.

Market volatility and fierce competition in the smartphone market (Apple's iPhone, Android phones supported by Google, and the once-mighty BlackBerry) weren't the only things testing Palm. Prior to its IPO, there had been a theft of Palm intellectual property that was never publicly reported, although it was reported to the FBI. Given Palm's valuation (for IPO purposes, at least) of roughly $25 billion, the theft of this core technology was classified by the FBI and was one of the top ten crimes on its plate at the time. The FBI never solved the crime, but 3Com's security department and legal team developed a good enough understanding of what happened, and were able to contain the damage as a result. Mark Michael and his legal team briefed the board of directors, but did not make any public disclosure. The intellectual property stolen was believed to a part of the operating system, and the theft was associated with a Chinese party. The IPO went off successfully regardless and the market was not bothered with this detail.

Mark Michael recalls other legal issues that confronted 3Com and Palm (while it was owned by 3Com) in this same timeframe:

> The two biggest or most strategic issues in the legal department over time were defense of securities class actions (stemming from the volatility of the stock—over a five-year period in the mid-to-late '90s, 3Com's stock price increased more than any other publicly traded stock in the U.S., from a split-adjusted $1.38 to $85 per share), and the development of intellectual property rights. On the class action front, we were targeted repeatedly and, counting actions in both state and federal court and cases filed directly against us or against companies we acquired and had to defend, there were over 15 securities class actions. We settled the first one for an amount less than the cost of further defense and settled the US Robotics case [for $259 million] but got dismissals in all the rest.
>
> The final case on my tenure involved exposure (pertaining to a shareholder class action suit, after the Palm IPO and the collapse of 3Com's stock in the early 2000s) in excess of a billion dollars and I got permission from Eric and the board to take it to trial. We retained David Boies[99] as trial counsel, with the firm of Wilson Sonsini as local counsel. Using a mock jury to test our chances, the mock jury rendered a plaintiff's verdict for hundreds of millions of dollars against us. Both Eric and Bruce Claflin put to me the question of settling, and I held out that it was premature and not the right thing to do. Three weeks before trial was to begin, we won a summary judgment motion and the case was dismissed.

---

99. David Boies also has been involved in U.S. v. Microsoft, Bush v. Gore, and, more recently and with some controversy, Harvey Weinstein against sexual abuse allegations.

The risky and fleeting nature of new technology—no matter how well-received or beloved it might be—is apparent in this graph of Palm's revenues across its history:

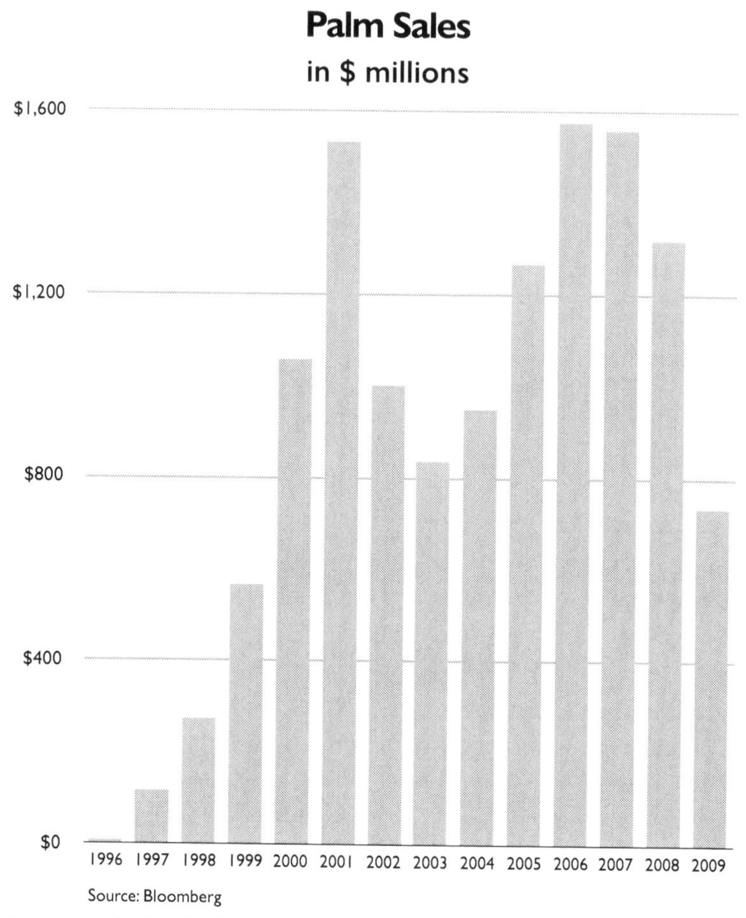

After Palm started off with a bang, it experienced some setbacks. The company was on the road to recovery—until Apple introduced the iPhone at the start of Palm's FY 2008, completely disrupting the market.

Chapter 18: Picking Up the Pieces

At the same time as Palm's IPO in March 2000, 3Com exited the market for high-end products like core routers and switches. The CoreBuilder Ethernet and ATM LAN Switches—gone. PathBuilder (a carrier-class ATM switch, that was being resold under an OEM agreement with Newbridge Networks) and NetBuilder WAN Routers were terminated. Extreme Networks took over the CoreBuilder products and customers, while PathBuilder and NetBuilder were transitioned to Motorola. 3Com's remaining systems focus was on Home and OfficeConnect product lines, SuperStack, NBX telephony, and Total Control product lines.

After this decision, 3Com would never gain back its momentum with large enterprise customers (at least not until the critical China strategy with Huawei played out after 2003). Sales director Brad Mandell also shared that while the exit of enterprise was somewhat a foregone conclusion, it was still damaging and left customers feeling like they were pushed to shop elsewhere. It gave the competition a great line: "Why buy 3Com? They may not be there in the long run with that product."

One 3Com executive remembered that, in a conversation with Cisco's John Chambers and Jim Richardson (a senior executive for enterprise products), Chambers and Richardson felt that 3Com was the only competitor they were scared of. From their perspective, 3Com was doing very well. They were shocked when 3Com leadership made the decision to exit enterprise, and felt that 3Com handed the remaining business to Cisco.

Anik Bose was working with the business units to help find some answers. This wasn't what he was originally hired to do, but it was urgently needed. In 2000, Anik Bose had joined 3Com as the VP of Corporate Development to work on M&A and partnerships. Previously, Anik ran the high technology strategy practice for Deloitte, which Eric had brought in for an e-commerce strategy project. When Edgar Masri decided to leave his Business Development role, he offered it to Anik. At that time, the company was operating at least six businesses: cable modems, carrier products, enterprise switching and routing, NICs, and Internet appliances.

In June 2000, Bruce Claflin, as COO, had tried to move 3Com into the market for smart consumer appliances by acquiring an Internet radio startup called Kerbango for $80 million along with a web surfing appliance called Audrey, which had a price tag of $500. Eric, Bruce, Jef Graham (spelled unusually as Jef), and Julie Shimer deserve due credit for taking risks, one of the core values of innovation. But based on a strategic assessment led by Anik, management decided to shut down the Internet appliance business, including Kerbango and Audrey.

Both were basically devices that made it simpler to get information from the web. Why did Audrey fail while, not that many years later, the iPad and today's household appliances like Amazon's Alexa thrive? One reason is that there weren't many broadband connections back then. And Audrey's user interface was too confusing, and some key bits of functionality (like its calendar) were weak. Today, one can simply ask, "Alexa, what's the weather for the next week?" and she will list off each day's forecast in a clear and useful way. Kerbango and Audrey weren't ready for prime time, and the company couldn't afford to wait for the market to catch up to the technology—not while hemorrhaging cash. Fast forward—in 2018, Lenovo announced the Smart Display, with Google Assistant, that is essentially what 3Com's Audrey was intended to be in 2001. Others point to Apple's iPad as well.

There was also steady innovation in the remaining traditional product lines. Jef Graham had helped build up the PCMCIA[100] cards for the laptops both before the USR acquisition and after. Doug ran the NIC business, and gave Jef responsibility for new products, including wireless, cable, and broadband initiatives. From USR, the division picked up the Megahertz's PCMCIA modem cards, and grew it from a $40 million business to ultimately $800 million. 3Com, Xircom, and Megahertz had been the key players, and after the merger, 3Com dominated the business. Ironically, 3Com had been about to come out with a Megahertz-killer product. Megahertz had been a leader in the modem card business, but were almost exclusively focused on the domestic market. Rich Redelfs recalled feeling that, "We don't want to buy these guys, we're going to put them out of business." Megahertz also had a combo card that was similar to one on the 3Com product roadmap.

Rich also recalled the development of a mini PCI card, to get Ethernet onto the laptops that were becoming popular in the late 1990s. With 50 percent share of the PCMCIA business for laptops, they negotiated secretly with the laptop vendors to build this little module on their motherboards, instead of using up the card in the slot of the laptop. They did this all without Intel finding out for some time. Kip Meacham, who came to 3Com as part of the Megahertz purchase, pulled off this coup.

In 1998, Doug Spreng retired, and Jef Graham assumed responsibility for the whole Network Connectivity Operations division from USR, the NICs, and mobile product business. He reorganized, and put Tom Werner in charge of the NIC business after VP Kef Kasdin, who had previously held senior roles in NSOps, left 3Com. Rich Redelfs was put in charge of the wireless and home networking business. At this point, Jef Graham was managing a $3 billion business with 7,000 people.

NICs were the cash cow that built 3Com and kept it afloat in hard times. But ultimately, they pushed the company into a classic "innovator's dilemma," when a disruptor gets disrupted. 3Com built the category for network interface cards (NICs) aka adapters, but was ultimately upstaged by Intel, which incorporated NIC functionality onto the motherboard via a chip, called a

100. The PCMCIA Card was used as an important expansion card to permit other devices to be attached to a laptop, including adapters, modems, and hard disks in the 1990s. This technology was obsoleted in the 2000s decade by external USB attached devices. See https://en.wikipedia.org/wiki/PC_Card for more information.

LAN on motherboard chip or LOM. By Q4 1998, Intel had a 51 percent market share versus 3Com's 34 percent in the 10/100 Ethernet NIC and LAN on motherboard (LOM) controllers in Q4 1998, and also led in Gigabit Ethernet shipments.[101] By commoditizing NICs into a chip or chipset, Intel also drove pricing down, something that worked for Intel's business model, but not 3Com's.

While Claflin was COO, a game of musical seats was being played out:

- Jef Graham was asked to assume a functional role, running product development for 3Com. Jef's group, Network Connectivity Operations (adapters, or what was remaining of them) was taken over by John McClelland under the umbrella "Global Operations".
- Edgar Masri, in charge of Enterprise Systems Division, assumed from Ron Sege's departure, also left the company and the product development for Enterprise was under Jef's Product development.
- USR was reorganized, and the modem business was spun out.
- The consumer business was being run by Julie Shimer (including the Audrey product based in Chicago).
- Gene Nelson was put in charge of the systems business.
- After Rich Redelfs left, Tom Werner took over the combination of the mobile and desktop businesses.
- Paul Fulton ran wireless, or what was left of it.

Later on, Claflin asked Jef Graham to shut down all the remaining unprofitable product lines. This included all new product lines, even those with wireless and broadband products. Jef Graham believed that 3Com would just end up with the maturing NIC card business, and that did not bode well for the future. Claflin and Jef could not reach an agreement, and Jef left in September 2000. The telco business had begun to slide in 2000, which precipitated Claflin's move to jettison these other disruptive technologies.

Jef Graham has since been a director on the board for Netgear (a leader in the wireless and broadband home networking markets) since 2003, and Julie Shimer later joined the Netgear board as well. After 3Com, Jef Graham became the CEO of Peribit Networks, a leader in WAN Optimization, later acquired by Juniper Networks in 2005.

Rich Redelfs left 3Com in 1999, and went on to be CEO of Atheros Communications. Atheros developed chipsets for wireless networking, or WiFi, and was ultimately acquired by Qualcomm. Jeff Abramowitz and Rich

101. Connor, Deni. "Intel's Network Equipment Gamble Pays Off". Network World. April 1999.

did a wonderful interview[102] for the Computer History Museum that provides insights into the politics and competitive dynamics that shaped the industry landscape of WiFi winners and losers. While still at 3Com, Abramowitz founded what became the WiFi Alliance,[103] a crucial industry group that helps promote the 802.11 standard (better known as WiFi), essentially making 3Com at least in name, the inventor of WiFi! Among other things, this alliance does interoperability testing for WiFi manufacturers. The original name for the group was the Wireless Ethernet Compatibility Alliance, as Ethernet was the core of the standard. There are now more WiFi devices in use than there are people on Earth, and more than half the Internet's traffic travels over WiFi networks. Yet 3Com, the company that coined the word WiFi, was nowhere to be seen in the WiFi business.

Life Imitates Art

There's a 2009 French movie called *Looking for Eric* about the escaping from the trials of modern life. Imagine that movie on a double bill with the 2003 comedy *Bruce Almighty* (2003) in which Jim Carrey's character learns how difficult it is to run the world. Together, these two movies about Eric and Bruce provide an apt portrait of what the company was like as 3Com's Eric Benhamou seeked some respite while 3Com's Bruce Claflin took on the difficult networking battles.

After leaving his day-to-day role as CEO, Eric stayed on as Chairman of 3Com. When we talked, he summarized what he thought were his main contributions:

> My contribution to 3Com was not so much inventing things. At best, I can take credit for spotting some trends, some inventions, and trying to anticipate what was around the corner. My best contributions to 3Com were helping to organize an extremely smart community of very smart people, not just engineering folks but people from all different aspects of the business profession, and getting them to perform at a high level and along the way creating shareholder value. I like to help organize, channel, focus, achieve, nurture along the way. But I know this is my vice: I did not seek the limelight. As soon as I had the opportunity to get out of the limelight, after ten years of running things, I took it. You cannot be CEO of a major technology company without being somewhat in the news and having to be

102. Redelfs, Rich. "Oral History of Jeff Abramowitz" Computer History Museum. December 2014.
103. You can find out more about the Wi-Fi Alliance at http://www.3comstory.com/tech.

> a spokesperson for the company on a regular basis. I relished the opportunity to go back to a less high-profile role, which is basically what I do today.

Eric also added some personal insights about the transitions that took place as Claflin took charge and the company made its fateful decision to streamline the business. He noted the sale of CommWorks later in 2003, (the subsidiary created after the USR acquisition for all of 3Com's carrier/telecom focused products, and part of the original golden goose Eric had pursued in the merger), to UTStarcom for $100 million in 2003, and their exit of the modem business earlier in 2000. According to Eric, the edges then evolved to be with low margin dumb devices according to carrier's strategy, without the software processing power they had hoped to promote. 3Com had envisioned smart edge devices. Eric clarified:

> Basically, they helped structure this part of the industry, the edge business, as a very unattractive business. Look, we had cable modems at some point, and we had DSL modems, so we had the technology to do this, but not in a way anywhere close to the margins we had on the NIC business. It was very hard for us to have a 30 percent gross margin business co-existing with a 60 percent gross margin business. Very hard to keep together. So, we decided to get out. This edge business became contaminated by the carriers, and then ultimately, the core business became less attractive because we didn't have a critical mass and the sales cycles were too long. There was a lot of pressure coming from the board, and from the unwieldy nature of the business itself.
>
> Two things happened that help explain the decisions being made around that time. One was the dot com bubble that burst, completely crushing valuations and slowing down the market in a very fundamental way. The second was that I decided to step down as CEO for purely personal reasons. I had been absent from my family for too long. I had an increasing level of family pressures to not continue as an absent father, so I had to step down. My wife went through a period of deep depression; my family spoke up at the time. I understood that I needed to wind down. I was confident that the board would anoint the COO as my successor, so wouldn't be leaving the company unattended, and I'll stay on the board.

These two events happened at the same time. I had already started the process of simplifying the company at the end of '99. I decided to spin off Palm as a separate company, so it wouldn't be an ongoing distraction. That also generated an immense amount of shareholder value. I pulled back from CoreBuilder 9000, and we got out of the large enterprise business. So, these two big things were off 3Com's plate, and I would have a successor, so the company can just keep moving in existing markets that we know.

What I did not anticipate is that—after I had turned over the reins to Bruce—that he would shrink the company so much more. I did not like him jettisoning so much, but I could not change that, because I had agreed to relinquish authority. You could not just hang around and undermine the CEO. So, I had to be as supportive as I could, acknowledging the fact that, from that point forward, he led the strategy. I would've done things very differently, but I could not have my cake and eat it too.

It was the right thing to pull out of the Audrey product at the time because the consumer market for networking and electronics gadgets was imploding. I mean, a few years later, Steve Jobs came up with a tablet. If you put an iPad next to the Audrey, you understand exactly what the concept was, so we were a few years ahead of our time. Similarly, with Kerbango... a few years later, Steve Jobs would come up with a portable version of it, he called it iPod. I think we understood technology, we understood the vision. But oftentimes, we pulled it in too early.

Bruce wanted two things, which I never insisted on. He wanted a simple business. Then, second, he demanded loyalty. He could not withstand disagreement or criticism. He expected people to "salute," meaning they were not to countermand his plans. This was a very different management culture. Frankly, I did not anticipate this when he was working for me, because he certainly managed upward very well. He managed me and the board very well. I never demanded that he be loyal to me. I always asked him to critique and propose different ideas, but certainly, when he was in the chair, that's not the way he worked. We were never friends, we were colleagues.

Another executive stated this about leadership decisions at the time:

> The team was starting to get tired. This had been a war, we'd all been working ridiculous hours for years. Perhaps if Finocchio had been made CEO, 3Com might have had a chance. When Debra Engel asked some folks, they said no—I think a lot of people were afraid of him. Yet Bruce Claflin came in, and was just all hat, no cattle.

When asked what he felt were the most critical elements to 3Com, Eric recognized that companies need different core values at different stages of their lives:

> I would describe it in three areas of competitive differentiation. The first one could be operational excellence, and you could put our supply chain excellence as an example of that. Another one would be technological differentiation, and excellence. The third is customer intimacy. I cannot think of any company which is excellent at all three. The best companies are excellent at two and are good on the third but not necessarily best in class. At 3Com we achieved excellence in technology, and a good proof point of that is that in the '90s we built by far the most impressive portfolio for patents in networking, over any other company by far. Much, much more than anybody else including Cisco and IBM.
>
> I think we also became operationally extremely good. We had great performance through all the traditional metrics. Customer intimacy is where we never achieved true excellence. Customer intimacy in our market required an extremely high-powered direct salesforce. Direct relationships with large enterprise customers is a must and while we made a very good effort at becoming better, we never really quite got there in terms of matching up competitively with Cisco.
>
> Cisco, on the other hand, had this trait. They were also excellent at operational management. Paradoxically, they were okay-to-good with technology, but not excellent. Oftentimes they were able to achieve extremely strong market share in some sectors without having the best technology. This is not an exception in the industry when this happens. Cisco paid close attention to what their customers, the CIOs buying their products, wanted. 3Com was, at heart, a technology-driven company, while Cisco was a customer-driven company.

While the story deserves more space,[104] Irfan Ali, the President of CommWorks, assisted in some high level comments regarding the carrier/telecom product space that helped fuel the critical USR acquisition:

> The [USR] Remote Access Server platform and CommWorks grew rapidly to $800 million in sales, with 60 percent gross margins. An IPO after Palm was in the works. We did have good customer intimacy; at one point, we had the top 20 telecom operators as customers. We also considered buying Sonus Networks, but the valuation was too high. We were developing a key product called Total Control 2000. It included capabilities for telecom operators to monetize on an IP based networking box for voice, data, fax video, wireline, and wireless. Then it fell apart. MCI, under Bernie Ebbers, was cooking its books and MCI was one of our largest customers. Along with a slowdown, the telecom market panicked. We had to take the drastic measures to downsize and weather the storm. Had the market stayed intact, CommWorks may have been able to go public, the product would have been successful, and the judgement of Eric regarding the USR acquisition different.

There is no doubt that external events—notably the collapse of the Internet startup bubble and economic recession—had their impact and made life harder for 3Com. But the real drivers of this phase of the company's story were things within the company's control: betting on certain technologies, product development decisions followed by product execution, effective mergers and acquisitions, and how Eric and other executives chose to run the company. (3Com executives unanimously felt that the executive team ran pretty successfully before the USR acquisition—and disastrously in the three years afterwards.) After Eric passed the torch officially, Claflin's tenure as CEO also had to contend with external economic factors, but his approach to leadership—not just externalities—was deemed by many a major contributor in what would happen in the next chapter of 3Com's life.

As the combination of acquisitions stacked up, 3Com never matched the Cisco direct selling efforts, and never got as close to their customers as Cisco did, even as it acquired more direct sales people and systems-focused product teams along the way. What it had was an increasingly teetering Jenga tower; it kept growing, but instead of becoming more powerful, it became more fragile and vulnerable. 3Com was betting the long game strategy on a systems strategy

104. Irfan's further perspective is covered in a break out section on our website 3comstory.com/tech.

to survive after the expected collapse of the adapter business, but it never fully evolved beyond the early years' lean-distribution sale channel model, with fewer customer touch points, and which relied on lower selling costs. This was the "original sin" from the Bridge acquisition, in which 3Com failed to invest in Bridge's internetworking products and also failed to compete with the high touch, direct sales approach that Cisco excelled at.

Eric is a great technologist and visionary, but Cisco gained early traction before Eric could take the reins by focusing on and owning the valuable parts of the market: the core of the network, products that were most "sticky" or most critical to their IT infrastructure. Yet this is still an over simplification. If 3Com had executed well on its product strategy throughout its middle decade, the trajectory no doubt would have been different.

2001: A Claflin Odyssey

"Those companies that have the resources, the skill, and the will to capitalize on convergence I think can have enormous changes in their market position. 3Com, to me, is one of those companies."[105]

—Bruce Claflin

While Eric was dealing earlier with the distraction of USR's channel stuffing (and related shareholder lawsuits and SEC actions), he was searching for someone to replace him after 10 years at the helm. He was looking for a leader with operational experience at a big company. That turned out to be Bruce Claflin, a former IBM and DEC executive with strong operations, sales, and marketing skills honed on IBM product lines. Claflin came on board in August 1998, as COO and President.

Debra Engel had interviewed Claflin, but was not supportive of his hire. Eric, in Debra's view, sometimes made hasty hiring decisions that didn't necessarily result in acquiring the best people. "Time was very important to him. An HR person usually says, 'Better to take longer and get the right person than to compromise at all for the sake of time.' Eric and I frequently disagreed on that front on more than one occasion. I could go through a number of executives that didn't last very long."

After leaving IBM, Claflin joined DEC; when the company was sold to Compaq, Claflin had a chance to take a sabbatical from work—but 3Com found him instead. This is how he ended up in the COO role first in 1998, an opportunity he felt would be interesting and new.

---

105. Duffy, Jim. "3Com gets ex-DEC exec". Network World. July 1998.

While Claflin was waiting to meet with Eric on his first day at 3Com, he noticed his own resume on the admin's desk. Curious, he took a peek. What he found, according to him, were glowing compliments, as well as an important point made by the recruiter: "Bruce Claflin has spent his entire career at two large, bureaucratic, East Coast companies. How will he fare in a fast-paced Silicon Valley environment?" How, indeed.

Bruce Claflin's tenure as CEO officially started in January of 2001. The former IBM and DEC executive from the East Coast "checked all the boxes" for the attributes that 3Com's board and Eric were looking for. Claflin was often described as a "sales and marketing" guy by many of the people interviewed and in trade publication articles from that time, but he also had plenty of experience managing domestic and international operations and P&L responsibility on his resume, including a successful run leading IBM's ThinkPad line in mobile computing.[106]

Claflin's primary success as CEO was ultimately, by nearly all accounts (including his own), embracing the company's successful re-entry to the enterprise market, via a joint venture with Huawei in China. Aside from that success, however, most of the 3Com alumni were openly critical of Claflin's leadership, and felt that he contributed to 3Com's decline. But as research and interviews progressed, it became clearer that the fate of 3Com's final decade was largely preordained well before Claflin became CEO, or even COO.

Claflin shared his perspective during several interviews on these topics and others—including his own self-assessment. When asked about his early years as CEO, Claflin sounded, as one executive suggested, like "someone that you listen to and think 'this guy would be great delivering the 6 p.m. news.'" His voice is not unlike that of NBC newsman Brian Williams. It was also noted by two executives close to Claflin that, like a newsman, Claflin appreciated well-scripted presentations.

Claflin's childhood was uneventful as far as events that might point to the trajectory of his future career. In trying to understand where he honed his instincts and skill, he pointed to his formative working years at IBM, "My early years at IBM were what really got me interested in business. I developed what I'll call my core values, and some of my capabilities there. Most of what I learned that was good I learned there, and the few things that were bad I learned there too."

When asked to describe some of his own leadership characteristics, he noted that hard conversations were best kept in private, and that praise and positive support are far more effective ways to motivate people, a philosophy he carried over from IBM. Claflin also talked about loyalty, noting how he

106. Lohr, Steve. "Notebooks May Hold Key to IBM's Revival". New York Times. June 1993.

remained loyal to his staff as long as they performed. IBM and DEC taught him to focus on results, goals, and achievements.

Claflin also shared an experience he had at IBM shadowing John Akers, who would later serve as IBM's president, CEO, and chairman of the board. He noticed that John would generally give the same pitch to different audiences, which didn't always turn out so well for him. Claflin's takeaway from this was that "a really good leader understands how to adapt to his environment." He said, "I tried to do that over time as I thought about how I was managing in a particular role." From statements like this one, it would appear Claflin characterized himself within what are called the "situational" and "contingent" behavioral theories of leadership.

This contrasted with what other 3Comers said in interviews. Many said that Claflin did not adapt at all to 3Com's culture, that he demanded loyalty, and that he did not invite debate or listen to the opinions of others. They also said that he didn't value the experience of the team that he took over. From their view, Claflin's top-down management style, his expectation that their executive team should fall in line, and the fact that he jettisoned even some of the most reasonable, technically-savvy, and networking-experienced executives, doesn't jibe with situational leading. Claflin's take was, "I welcomed debate, but not endless debate. And what I did demand was that once a decision was made it would be supported 100 percent. I don't call that loyalty. I call it leadership."

Many of the long-time 3Com execs that didn't stay on after Claflin's promotion felt that Claflin wasn't merely not making any attempt at all to hang on to them but that, in fact, they were essentially chased out of the company. Claflin's take is a bit different, believing that any initial action he'd taken to let someone go contributed to an overall air of nervousness among the management team, which created a downward spiral of anxiety that prompted some folks to leave. Claflin also felt that it was only natural for people to leave, faced with the arrival of a new CEO who would want to fill their own cabinet. He felt that there was finger pointing at Eric, they didn't buy into the strategy, and there was endemic stress from lack of growth and profitability.

Claflin had certain clear expectations of the company he was joining:

> I had an image that it was going to be a group of meat eaters: fast moving, decision-oriented, really just maniacally focused on business performance, as opposed to what had been the position of large bureaucratic East Coast companies. And frankly, it was the exact opposite. My initial observation was that it was like a United Nations meeting. Everybody thought they had a vote, anybody could veto anything. Frankly I was stunned at what I believed was a lack of a "fast pace" discipline.

This appears to be fair criticism. While Eric did expect his executives to behave like rational adults, ExecCom had plenty of food fights and turf battles, before and after the USR acquisition.

Yet the question remains for any new CEO—how do you lead a technological company, without mentors or execs at his side, steeped in the company knowledge and products? Claflin may have rightfully had a view to swap out execs, and make his own mark over time. But the trick he missed at the start may have been overlooking the usefulness of the guidance from his team, which destroyed morale, throwing away resources at his disposal. The typical new leader mantra of "observe first, listen, ask questions, then proceed," can help shape, deepen, and sharpen a CEO's vision.

One executive suggested, "You could imagine, that he could have bought astonishing levels of loyalty, for whatever was the deemed right amount of time, by asking for what they could do together, a win-win in the transition, so he could get continuity. ExecCom did represent a set of resources for him, he should have used—in fact, exploited may be a better word."

Claflin had another interesting insight about what he discovered when he joined. He noted that he was brought in to help assimilate a large and complex $6 billion operation, and augment Eric, who was viewed as more of a strategist. He perceived a schism between the desire to integrate the disparate parts of the company versus the notion of letting this conglomeration of operating units (none of them clearly dominating in their markets) run more like a federation of separate businesses.

The strategy of being a systems company was never going to work, Claflin realized in hindsight, as 3Com had become a diverse portfolio of businesses. He stated that even in later board meetings, the board wasn't clear about whether the company was a portfolio or a company in the process of evolving into an integrated systems company.

In addition to this existential ambiguity, Claflin also became CEO at a ridiculously bad time. He recounted what was happening at the start of 2001. The new CFO and Claflin had determined that at the rate 3Com was losing money, it could be out of business in 16 months. Per Claflin, "The context was we had to preserve cash, we could not be making big investments that were gonna take years to pay off."

"Bruce was the CEO, and said, 'OK, let's go and turn the remaining businesses around,' but guess what? It's one thing being a strong sales and marketing general manager, who can grow a stable business. It's another thing trying to turn around a sinking ship."

Anik Bose, VP of Business Development

Death and Destruction

Claflin had the unenviable task of overseeing the greatest destruction of cash, net worth, and shareholder valuation in 3Com's history during his first year as CEO.

3Com's $4.3 billion sales in FY 2000 plummeted to $2.4 billion in FY 2001.[107] In FY 2000, 3Com's cash and investments had risen to $3.1 billion, an increase of $1.4 billion from the end of FY 1999. (That included $1 billion from positive operating activities, along with $800 million from its gains on investments such as Palm stock, offset by $400 million in capital investments and other investing and financing activities.) But one year later, by the end of FY 2001, roughly half of 3Com's cash was gone, having fallen to $1.6 billion. Operating activities had burned through $1.1 billion, and the company wrote down its investments by an additional $500 million, with more write-offs ($450 million for restructuring charges and $229 million for acquisitions) to come over the next few years.

The board, Claflin, and the executive team were dealing with issues that should have been addressed sooner than they were. They could not be running all these businesses with 14,000 employees and negative cash flow. The company needed to quickly figure out what to stay in and what to get out of. Gwen McDonald, the VP of HR, who supported Doug Spreng in the adapter division throughout the 1990s and later oversaw HR under Bruce Claflin, said, "This didn't start with Bruce, but it just continued… really, what was 3Com's mission? We lost our purpose along the way. We tried a lot of different experiments with varying returns, but at some point, particularly under Claflin, we just did not know."

In 2000, Claflin recruited George Everhart, a PC sales executive from the likes of Apple and Fujitsu, to run 3Com Sales, with the hopes that he might later become COO. After the downturn, this possibility evaporated. Upon arriving in mid-2000, Everhart quickly realized that trying to turn things around for 3Com would be tough going, especially with so many "A" players leaving, and the formidable challenge of competing in Cisco's shadow. Everhart did spot a few reasons for optimism at the time. "Matthew Kapp in Asia Pacific had that part of the world pretty excited. And Raphael Fernandez, in Latin America was one of the few places in the world (along with South Korea and Brazil) where we had outgained Cisco in market share the prior year. DSL modems were doing well."

But aside from those bright spots, Everhart felt there were deeper challenges. Bruce Claflin, he felt, was not a technologist, and didn't claim to be. He also didn't seem to roll up his sleeves as much as Eric had in the years earlier. This perception of Claflin negatively impacted the troops, in Everhart's view,

---

107. As a Director of Accounting and Reporting, I was dealing with similar growth slowdown problems at telecommunications giant JDS Uniphase (JDSU) in the same year and wrote $50 billion worth of "goodwill" off JDSU's balance sheet as our enterprise value crashed to earth on the weakness of orders from our customers (such as Nortel and Alcatel-Lucent).

especially those that had a technological background and didn't see Claflin referencing or taking advice from the technical community.

Another pervasive issue was the push to improve cash flow by reducing inventory and tightening lead times. On the positive side, this did help the financials. But it came at the expense of making the sales cycle tougher. Salespeople were worried they would not get the products they needed when they needed them. This created the temptation for them to inflate forecasts to ensure they would be able to deliver on a sale.

Everhart recapped, "If we had a clearer direction of where the company really was headed, in a positive way, it would have been less of an issue. It's like a football team. When they're winning, the fact that a couple of defensive schemes aren't working as well, people look past it. If you're losing, everything that happens is bad."

Broadcom Broadsides 3Com

Around this time, Claflin also began disposing of the USR cable modem business, which was losing millions of dollars a year and was spun off in June 2000. 3Com completed selling off its CommWorks subsidiary to UTStarcom, in exchange for $100 million in cash.

> **"The year 2001 has been really lousy. I think we are all just looking forward to turning the page on a lousy year."**
>
> *Eric Benhamou*

This was a sad finale to the mostly unfortunate USR story. Eric had advocated for the $6.6 billion USR purchase because of its modems and, more importantly, its carrier class products that would enable 3Com to compete in carrier switching and higher-level protocol technologies against the likes of Cisco. The valuation turned out to be overhyped, modems were obsoleted, and the carrier products were sold off.

In the wake of this dire financial picture, no ideas to generate revenues or cut costs (or, ideally, both) were off the table. Even the most sacred of sacred cash cows—the adapter business—found itself on the chopping block. Could 3Com adapt to life without adapters? There also were efforts to create a side business making chips that would provide a "LAN on motherboard" (or LOM) to compete against the specter of Intel or others slipping networking functionality into CPUs or onto computer motherboards. But making chips wasn't in the company's DNA. Getting out of the adapter card business was inevitable—it was just a question of "when?"

How did 3Com find itself at the point of jettisoning its original golden goose, the NIC business? According to Claflin and others interviewed, the NIC business had started to erode much earlier with the advent of 10BASE-T technologies that used twisted-pair cabling (instead of the more unwieldly coax cable used in the original Ethernet connections). And by 1995, Intel had made significant inroads into the NIC business, and 3Com's high price point exposed them when competitors like Intel undercut their prices in 1997.[108] 3Com also suffered from a deeply entrenched preference to sell cards, not chips, or to pivot to something entirely different (network management? Services? VoIP? Home networking?). In other words, 3Com had a classic case of the innovator's dilemma: after disrupting the marketplace via Etherlink III, it failed to notice how the market they had reshaped was once again being disrupted.

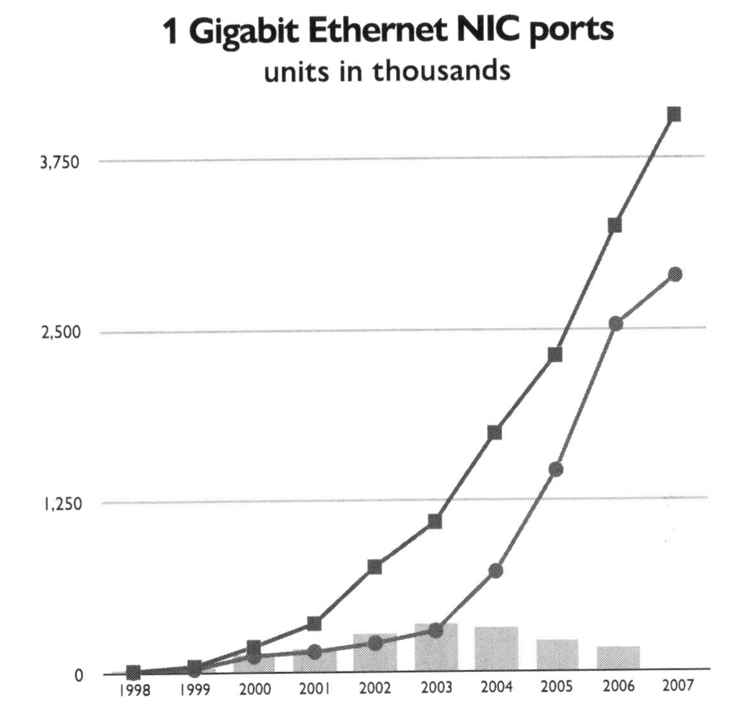

3Com ultimately conceded the NIC business to Intel (and others) after 2000.

And finally, and perhaps most importantly, the NIC business was being dwarfed by the LAN on motherboard chip business by Broadcom and others. Broadcom had carved out a successful strategy of supplying chips to 3Com (knocking Lucent

108. According to Ali Khan, a Marvell Product Managerment Executive and 3Com alum "as an example of the decline in pricing, a Gigabit Ethernet server adapter was $1,695 in 1997. In today's market, a LOM chip with even more functionality on it would be less than 50 cents."

out of the 3Com account) and also developing and producing their own NIC chips to sell to their own accounts. Broadcom was able to entice OEM motherboard customers like Dell and Compaq to buy from them instead of 3Com as technology moved to 1 gigabit Ethernet speeds. Meanwhile, Intel had absorbed the 100 megabit chip into their Southbridge chipset on the motherboard.

As Tom Werner explained, 3Com wasn't simply sitting still and inviting others to eat their lunch, "While we waited for Gigabit Ethernet to come along, we added an ARM processor to the 100 megabit adapter that enhanced security, and we were getting some traction." He also noted that 3Com had made some attempts with SMC Networks (now a subsidiary of the Taiwanese company Accton Technology Corporation, and no relation to the Japanese company with a similar name) to build chips on motherboards like Intel's, but by then Broadcom had developed a complete chip solution and was aggressively taking over 3Com's customers. Bruce Tolley, a 3Com marketing alum, also pointed out that 3Com's options for LAN on motherboard technology wasn't limited to Broadcom: Marvell—which later offered their technology to 3Com, may have offered another alternative at the time, and 3Com may have had rights to Alteon's intellectual property that would have helped block Broadcom's rights to the technology.[109]

By the time Gigabit Ethernet came along, Broadcom and 3Com still had an alliance of sorts whereby both contributed technology to various products. But Broadcom had maneuvered its way into the driver's seat for that partnership, since the agreement allowed them to sell directly to any OEM. Broadcom now had the IP and know-how to make those kinds of sales, and their customers were motivated to eliminate the middleman (namely 3Com). It doesn't appear that 3Com put up much of a fight, and sales of Gigabit Ethernet adapters went to Intel and others, while the sales of LOM chips to OEMs like Dell were taken over by Broadcom.[110]

Jeff Thermond worked on the potential sale of 3Com's adapter business to Broadcom, but from the buyer's side at Broadcom, and thinks the company wasn't that far off from a sale. Former 3Comer Jeff Thermond had left 3Com in 1997. He had landed at Broadcom when he sold his company Epigram (a leader in broadband home networking chips) to Broadcom for $500 million

109. For a more detailed explanation of the PHY and MAC adapter technology, visit 3comstory.com/tech.

110. The antics of Broadcom's larger-than-life Co-Founder Henry Nicholas—who, among other things, built an actual underground lair beneath his more conventional Laguna Hills mansion—make an interesting backdrop for this episode; you can read about him in the 2008 Vanity Fair article at https://www.vanityfair.com/news/2008/11/nicholas200811

in 1999, equal to 5 percent of Broadcom's stock when it closed in 1999). Jeff noted that Dell had approached Broadcom saying:

> We know you're in the lab manufacturing the chip, and we know that a significant amount of this design is yours. When we have problems, they're usually physical layer problems, and the 3Com engineers didn't do any of that, so they can't answer our questions. They don't let us get through to you directly, and we actually need some help here. Frankly, we'd save a bunch of money going direct with you.

Around the same time—in 2001—3Com was trying to sell off the adapter business. Broadcom wasn't really interested in the adapter business per se, but they were very interested in the underlying patents. At the time, Intel was suing Broadcom (for half a billion dollars) for infringing on Ethernet patents. Understandably, this was driving Broadcom crazy. From his new perch at Broadcom, Jeff observed, "We were racking up legal expenses out the wazoo. I was also considering buying Motorola's soft modem group, a similar business. I wanted the patents because I was getting sued by a bunch of people." Bringing a sizable portfolio of intellectual property (IP) into Broadcom's corner would be an enormous help in staving off the legal challenges from Intel and others.

Jeff Thermond took the place of Marty Colombatto (Broadcom's combative Ethernet executive) in the negotiations to acquire 3Com's NIC business, Marty was not liked at all by 3Com, and Jeff had prior history as a 3Com employee, so the negotiations appeared to get back on track. But the deal fell apart at the eleventh hour according to Jeff. "I was prepared to offer about $200 million. I said I'll buy it, but you've got to give me a pretty fair whack at those patents coming over. Their position was 'Wait a minute, these patents apply to our hub business, our switch business, our router business …' My response was 'That's your problem, not mine.' It may be easy to build a company up through acquisitions, but not everybody's IP strategy is built to be taken apart."

3Com concluded that there was no deal. Bruce Claflin offered this clarification: "The price for the adapter business had started at $400 million prior to Broadcom's due diligence, after which it dropped to $200 million. The board and I felt we would benefit more by running it for cash for a while longer, so we took a pass. We believe that we did generate more cash for a longer period of time, in excess of what we were offered by Broadcom."

To borrow Geoffrey Moore's analogy, the adapter business was not ultimately a technological gorilla that dominates the standards, thereby controlling the value chain (the way Cisco dominates the enterprise network space today). The adapter business was one led by a king (market leaders that do not have proprietary architectural control over their category) and kings can get knocked off their thrones.[111] By 2003, there were more than four billion 3Com installed connections, including adapters and ports. While the adapters were 3Com's original heart and soul, accounting for 53 percent of all business in FY 1995, by 2003 the division contributed only 16 percent of total revenue, and that number continued to fall rapidly.[112] As Bruce Tolley summarized, "Fundamentally, 3Com's failure in adapters was neither technical, nor manufacturing related, but involved the strategic decisions or lack thereof around what to invest in."

"In early 1996, Jeff Thermond and I went down to Orange County to meet with Broadcom—specifically Henry Nicholas and Henry Samueli. Mostly to discuss about cable modem PHY. I was still pretty new to all of this. After the meeting when I realized the full portfolio of what BDCM had I suggested to Jeff that 3Com should own these guys. Jeff said that wasn't crazy. When Nick (Henry Nicholas) called to follow up I told him 3Com wanted to buy the company. He threw out a huge number. I calmly told him I thought we could get that done. He was speechless for a few seconds and then quickly said Broadcom was not for sale."

Steve Foster, 3Com Business Development Executive

The Why about WiFi

If 3Com was blindsided by the rapid demise of the adapter business, was it equally blind to the growth potential in other markets? In particular, why didn't 3Com have products in the booming WiFi markets, where products like home routers and access points were taking off? Per Bruce Claflin:

> WiFi, obviously, was very exciting. It was gonna play a role both in the home and in business. But the early WiFi standards—B, A, and G—were fairly slow. There were some reliability issues about signal transmission, and WiFi performance, depending on a lot of things (like the walls in a building and the placement of devices) that we couldn't control. Still, it was obvious that WiFi devices were gonna be important. But we owned absolutely no critical IP in the wireless

---

111. Moore, Geoffrey A., Paul Johnson, and Tom Kippola. "The Gorilla Game". Harper Business. 1999.
112. Shore, Joel. "The 3Com Saga: One-time Industry Pillar Hits 25". Network World. April 2004.

Survival in the Land of the Giants

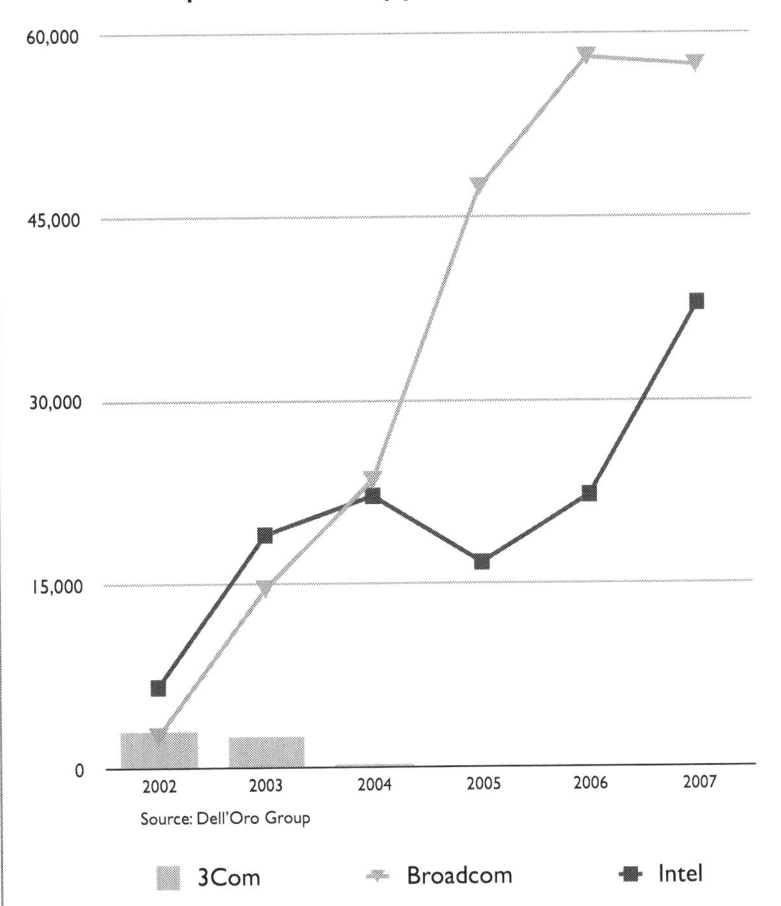

Despite Intel's vast resources and market clout, Broadcom's success in the LOM chip market was proof that there was always a way to best the industry Goliaths. (In Broadcom's case, that included a take-no-prisoners approach to business.)

> stack. None. So if we were gonna be in that business, it was gonna be as a systems integration sales and marketing company.

Selling WiFi products for the home, he noted, would limit the profit potential, as the consumer electronics channels ate into margins and suffered from high return rates. And companies like Netgear and, a little later, Cisco, had already established strong positions in this emerging market. But 3Com had high volume manufacturing, great engineering, and USR's modem background. Was this raising a white flag and ceding the battle? Claflin continued:

> Meanwhile, our consumer products, principally modems, were collapsing, based on aging technology. So here we were with this dilemma. We knew there was a growth opportunity. We controlled no intellectual property. We had no cost advantage. It was largely a packaging and channels play, and the products that we had in the channel were evaporating based on movement to broadband. So that's why we didn't go after it in a big way.

3Com's original foray into the WiFi space—via the acquisition of Pacifiic Monolithics in the 1990s—may have tainted the potential of WiFi for some 3Comers. Rich Redelfs recalled, ""Pacific Monolithics was developing a proprietary 10 Mbps WLAN product at a time when the prevailing IEEE 802.11 standard only supported 1 Mbps speeds. But the product never quite worked and development continued to fall behind even as the 11 Mbps IEEE 802.11b standard arrived on the scene. 3Com pulled the plug on further development and then partnered with Intersil for chips and Symbol Technologies for systems, to create a IEEE 802.11b wireless offering for enterprise WLANs, but was not ultimately successful with WiFi."

"This was during the tech bubble. If your stock wasn't going up by some outrageous amount each quarter, you had people quitting, to go mine the technology gold rush. We did keep running the [WiFi] business for a time. I think the WiFi business was more a matter of attrition combined with the parent 3Com not doing well. A combination of the gold rush exodus combined with Bruce's cost cutting."

Tom Werner, 3Com Executive

The CTO Role

Sometimes, a strong CTO can solve real problems in addition to setting the technological vision. In the late 1990s, 3Com had a Gigabit Ethernet adapter that IBM had bought in the millions of units. It turned out there was a defect in the adapter discovered in 2002, and it didn't fully comply with the Ethernet spec. The product had a defect that only kicked in when traffic exceeded a certain threshold. After going up and down the issue with the engineering teams, there was not a single engineer that could figure out what was causing the problem (let alone how to fix it). They ultimately called Paul Sherer, 3Com's former CTO, who had recently left the company. In true Paul fashion, he came in over the weekend and fixed the bug.

This is one example of how 3Com had lost its way—its engineering resources had been so seriously gutted that its own development team couldn't make a noncompliant adapter work. This was a complete embarrassment.

Paul Sherer was the last person to hold the title of 3Com CTO. In 1995, he was given the charter to create the Technology Development Center (TDC), which helped with product innovations that could be used by the product divisions. Paul succeeded John Hart as CTO in 2000, just as Claflin was taking over. Debora Chase, my wife and Paul's former executive assistant, reflected on the elimination of the CTO role, "Paul did not have a conduit to the business units to sell his ideas. There was no one to evangelize them inside the business units." This was compounded by lack of strong ExecCom support. And Paul's good natured personality didn't help drive action in the divisions. Yet imagine, spending millions of dollars on Paul's engineers and the TDC and their innovations, yet those ideas were basically ignored by the product divisions.

Paul Sherer, frustrated in his dealings with Claflin, the inventor that created vast wealth for 3Com with the EtherLink III amongst many inventions, left in August 2001, after 17 years with 3Com, and Claflin eliminated the CTO role.

Can an astute CTO provide the necessary guidance in navigating through rapidly evolving technology markets? Claflin's take was that they were not involved in the product decisions in the business units, and more useful in front of customers or showcasing ideas at conferences. But he did concede that the business units could have benefited more from thinking outside the box and work with innovations outside their units.

Claflin's take was that the business units were iterating on products that would allow them to continue their revenue streams, not so interested in listening to the TDC or CTO talk about the next new promising technology, which is the classic innovator's dilemma—stay with your runners, vote against things that could hurt the larger ship, such as smaller niche products requiring patience and investment. No wonder the TDC or CTO would not have an easy time selling a breakthrough idea to the units. But Claflin did go on to point out what he viewed were some of the differences between 3Com and Cisco while talking about innovation and technology:

> 3Com tended to sell lower ASP products, they tended to be kind of hot products at the time, but they didn't have the stickiness that you would get on a systems type of business. And many of them became commoditized pretty quickly.

This was supported by the dilemma faced earlier in the 1990s during Eric's tenure when Cisco was wrapping its arms and routers around the CIO and mastering the art of building their direct sales channel. But perhaps hiring Claflin, who was familiar with mature product markets like PCs and laptops, may not have been the best fit with the fast evolution and innovation of Ethernet and networking in the late 1990s. As Satya Nadella puts it in his new book *Hit Refresh*, "Our industry does not respect tradition. What it respects is innovation."

Peter Wang, a Technology Director hired by Paul into the TDC in the mid-1990s, shared his view of the technology innovation at the time:

> It's not only the passion from the top, but also the conviction and the strong corporate wide synergy and execution that will win the day for any company, not just in the startup. Bill Gates' changing the massive Microsoft's strategic direction to the web in the mid-'90s was the most vivid example of this. If Microsoft hadn't handed the reins to Satya Nadella and changed its strategic focus to the cloud and AI, with the necessary backing behind the shift, it would be losing its relevance today. It is 20/20 hindsight, but from a learning perspective, one has to be critical and ask, "What has the 3Com executive leadership and the board failed at, beyond what you have written?"

In discussions with Claflin, his take was that technology matters most in helping create a great business, and he was most passionate about creating a great business, with technology as a means to that end. Perhaps the alignment of 3Com's product stars were simply not fixable by Claflin, but others still held him responsible for much of the pain.

Quid Pro Status Quo

In her book *Closing the Innovation Gap*, Judy Estrin offers a cautionary warning about the impact of the Internet bubble that peaked—and imploded—in 2000 on Silicon Valley. She notes a "growing concern about the state of innovation in the technology industry. Silicon Valley had changed. The excitement of solving interesting problems by creating new technology was fading, and immediate financial returns were taking precedence over the mission of building lasting companies." Aptly put as 3Com competed for mindshare and multiples to garner Wall Street's attention.

Judy's book also discusses patience, which she considers one of the core values of innovation. She cites 3Com's decisions to abandon strategically critical markets based on short-term financial motivations. "'So we pulled out of the enterprise business, at a time when we could no longer satisfy investors' inflated expectations, already distorted by the dot com bubble' recalled Eric Benhamou, 'thus jettisoning a foundation built over 5 years.'"[113]

Judy Estrin also noted in her book that the following core values must be associated with innovation: questioning, risk taking, openness, patience, and trust. Without these, the capacity for change is limited. Given all the turmoil, and for all the reasons and explanations covered herein, during Claflin's tenure, products were jettisoned to protect the balance sheet and cash position.

As an example, 3Com exited broadband and wireless products for the home as those markets were taking off, although various reasons for their exit have been offered up. Netgear stepped in, and today sells huge volumes of WiFi routers and other networking gear. Netgear grew 267 percent to $299 million in sales from 1999 to 2003, and is currently a $1.4+ billion revenue business. Patrick Lo, the CEO of Netgear, was quoted in 2009, "With their strong salesforces, Cisco and HP are too well-established in the enterprise to be unseated. Netgear's strategy in the U.S. has been to target 3Com accounts, and Netgear's rise mirrors 3Com's decline."[114]

IBM, Cisco, and other companies eventually found this existential challenge of the necessity of reinventing themselves. They all face the innovator's dilemma—needing to sacrifice their profitable business to reinvent themselves into something else—even if that means pursuing a smaller niche market that they hope will later grow and blossom. One more example: While 3Com ultimately tried to enter the security realm with their purchase of Tipping Point, the security issues existed much earlier. Imagine if 3Com had pursued security in a much more concerted way, using its Israeli talent in security for example, along with its vast networking knowhow.

Paul Rudnick, VP of Engineering from 1999 to 2002, indicated that security had been looked at in the adapter arena, but it was viewed more as a niche (the innovator's dilemma). In his opinion, 3Com was in a unique position at

---

113. "Closing the Innovation Gap". Page 21.
114. Macpherson, Sholto. "Netgear reveals its full ambitions". CRN. March 2009.

the edge of the network to create secure NICs, perhaps preventing the later hacks of the hundreds of million customer accounts such as Home Depot, Equifax, and others. The technology was looked at with Secure Computing Inc. (later bought by McAfee). Paul weighed in, "This was one of the decisions so poorly made. The loss of gigabit adapter technology to Broadcom was another 'dumb' decision. Claflin had shut down three design efforts to create such a component set. The decision to outsource manufacturing was third. Finally, the election to fire all engineering, 4,400 employees, with marketing's encouragement, and outsource to Huawei was just stupid. All 'tribal' knowledge was lost." Today, security is a multibillion-dollar market, led by names like Check Point, Palo Alto Networks, Symantec, Trend Micro (who bought Tipping Point), Fortinet, and more.

Consider the business trajectory of one of today's most highly respected innovators: Amazon. That company began as a bookseller, with the relatively minor wrinkle of conducting their sales online. Amazon later commercialized its own cloud infrastructure and turned that into a separate business—Amazon Web Services—that generated over $7.4 billion in 2018, and is projected to become its own $18 billion business. The company has become a major player in producing original content to feed into the media streaming devices it makes. And to help improve the user experience of the voice-recognition that powers those devices, Amazon announced in 2018 that it would be making its own AI-fueled chips.

"You need to love your product. You need to love it because that's the difference between winning and losing. You need that passion. Bruce is very competent and very smooth. I've known a number of people who've worked with Bruce on boards, and they are quite complimentary. But if you run a business in the valley in the tech industry, you've got to love it. You've got to want to be there. It's hard to get too excited about it sometimes, but what it does is very exciting. And so, you have to love what you do. You have to love the products. You have to love the people. You have to be engaged in part of this community. Obviously senior executives don't necessarily hang out with everybody all the time, but there is a community. There is a network, and you need to be part of that."

3Com Executive (who wished to remain anonymous)

If a bookseller can wind up in the cloud computing and chip business, it's not hard to imagine that 3Com might also have found solutions to reinvent themselves. Of course, it's easy to speculate about such things in hindsight. At the time, it would have been tough to assess whether it was a good idea or a distraction—"Hmm, do I compete with Cisco in enterprise, Netgear in SMB, or shall I take on Broadcom and Intel chips today? Perhaps try to check Check Point?" But coming up with outside-the-box ideas and wrestling the pros and cons to the ground goes hand in hand with innovation.

¡Ay, Charisma!

Bruce Claflin represents one version of the charismatic theory of leadership; many examples are on display daily for us to see in American society and are easy to spot and all of 3Com's leaders had their own charismatic traits in carrying out their goals and vision of the future that followers could buy into. Bruce for example appears to have relied on his charisma and strength in communicating effectively upwards to the board, but didn't succeed as well in leading those below him. Claflin himself, points to the two successful things he contributed to as a top leader at the company for eight years (six of those as CEO): protecting the company's cash, and starting H3C.

Claflin's focus on metrics, supply chain efficiency, cash flow, and the like are respectable hard-edge values but also came at the expense of soft-edge values, like trust and teamwork. As Rich Karlgaard aptly said in his book *The Soft Edge*, "Great, enduring organizations are masters at both the hard and soft edges."[115]

Perhaps Eric and the board approached the hiring of Claflin (and similarly Carl Yankowski later at Palm) as a conscious effort to hire a "best athlete," a corporate jack-of-all-trades, business-suit candidate, with an operational, sales and marketing background. A recent *Harvard Business Review* article suggests that hiring is most successful when it considers the particular contextual leadership challenges. Rather than simply take on a generic workhorse leader, companies should assess the attributes and experience of the challenges of the position and pick the right thoroughbred accordingly. Considered in this light, the Bruce Claflin hired to be COO of a $6 billion company in 1998, looked quite different from the executive candidate elevated to CEO in 2001, then hired to turn around a technology-rich, product-diverse, and rapidly collapsing Silicon Valley company.[116] Nevertheless, he

---

115. Karlgaard, Richard. "The Soft Edge: Where Great Companies Find Lasting Success." Jossey-Bass, 2014.

116. Harvard Business Review, September-October 2017, "When Hiring Execs, Context Matters Most," page 20. The researchers consider a wide variety of contextual challenges—a leader may be needed for growth (whether through innovation or by focusing on cost competitiveness), geographic expansion, managing a broad portfolio of products and services, leading through M&A, transforming a conflicted culture, or leading global or cross-cultural teams.

was a known quantity, a leopard does not change his spots, and the board and Eric were comfortable and promoted him to CEO.

Bruce Reviews the Boss

Did Claflin have any regrets, or things that he would have approached differently—in his leadership, with his executives etc.? He admitted to having regrets, hiring issues, turning over people. But he pointed out the collapsing markets and need for protecting cash, and managing a stressed organization.

Claflin came into 3Com with a strong resume. But almost immediately, he was expected to shake up the product mix, overhaul operations, and ultimately restructure the company. His skills in communicating effectively to customers and investors weren't necessarily going to translate into a knack for transforming an organization. Did Bruce ever feel "miscast" as 3Com's CEO?:

> It's a very good question, and I remember talking with one of our directors as I was nearing the end of my time at the company, and he was trying to give me a compliment—although it didn't come across that way 100 percent. He said, "Gee, when we hired you we thought what we needed and what we were getting was this incredible operations guy, and what we got instead was a brilliant strategist." And then I looked at him and then I said, "But not as strong operationally?" And he sort of looked at his shoes and said, "Well, you know," but that's what he meant. I think there's some truth to that.
>
> On my last day with the company, we were having a board meeting with my senior team. The new CEO Scott Murray was standing there with me; we had spent a couple of days just talking through the transition. I was saying goodbye to the team, and said, "At times like this, you tend to be reflective and ask yourself major questions," and I said, "So I've asked myself what did I do here? Did I do anything that was any good? I concluded that I did one and a half things right." And they started laughing because, you know, after five years as CEO you did one and a half things?
>
> I said that very early on after joining the company, while I was still COO, I recognized that we had the ability to extract a lot of cash out of the company, particularly out of its supply chain. At that time, by the way, when I joined, our cash-to-cash cycle was about minus 90

> days, which basically means that you're a bank. You're financing your customers and your inventory. By the time I left we were positive 17 days, and it generated about $500 to $600 million of cash, one time.
>
> So, five years later I said, "I think one of the things I did that was right was early on recognize that we could extract cash from the business, and that cash ended up allowing us to survive when the downturn came. I really believe we would be gone had we not done that. And so that's the one thing I did right."
>
> So, what's the 'half' that I did right? I said, "I think we now have options for growth.' At that time, we had two big plays open to us. One was the Huawei venture, and the second was Tipping Point. I said, "Now I don't know how they'll pan out, you all are gonna figure that out after I'm gone, but they are options for growth that we didn't have before."
>
> I still feel that way today. You make thousands of decisions that are not breathtaking, and some are good and some are bad. But those would be the things that really made a difference.

When asked about whether he thought he should have deliberated more with the team in place, and been more receptive to new ideas, Claflin weighed in:

> If I had to criticize me, I'd say I wasn't enough of a "command and control" CEO. These were, after all, very desperate times we were in. Good companies were struggling, struggling companies were dying, and I wish now in hindsight I had been more decisive. I allowed things to go on too long. I should have moved much faster to shut down Kerbango, Audrey, and the whole consumer business—including the analog modem business.

Regarding the executives that left after he took over, Claflin noted that it was a tough sell and felt his attrition was consistent with the market, but agreed that high turnover at the higher levels did not help the company:

> I'm not surprised that some left quickly, and others stayed quite a while. I hired a number of people in key positions, starting when I was CEO. By then, most of the team that was going to leave—that had worked directly for Eric—had left. I was hiring now for the first

> time as a CEO bringing in new blood and, frankly, a lot of the execs turned over at my request. An executive role in a troubled, struggling company can be a really difficult job.

Bruce was asked about some other decisions that had their share of detractors. One was moving company headquarters to Massachusetts in 2003. Cutting to the chase, his answer was largely based on the fact that there were scant operations here in his view (no more Palm, manufacturing, data center, etc.), and the enterprise business being run out of Marlborough was most important. While not wanting to do it purely for cultural reasons, he felt that shaking it up a bit wasn't a bad idea. He also strongly disagreed with any illusion that he was an East Coast guy wanting to get back home.

Despite Claflin's rationale that convinced the board to move 3Com east, all those interviewed for this book (and virtually all West Coast alumni) felt that this was a completely wrong decision. They wondered how Claflin so quickly dismissed 3Com's deep roots in Silicon Valley, and its strong symbolism in the area as a company of great importance in the local landscape, and wondered aloud if earlier decisions made were in furtherance of his moving the headquarters east.

While 3Com had a campus on the East Coast, its center of gravity was always in Silicon Valley. The Bay Area was not just the center of gravity for the company, but of the entire industry (Cisco, Bay Networks, etc.). As an East Coast fellow, everyone felt that Bruce had some inherent bias, and that he never really felt comfortable here.

"It's hard to focus on culture when you have such big problems. In those days, when you have to go from 12,000 employees to 4,000 employees—guess what? Culture falls lower on the priority list."

Anik Bose, VP Business Development

Reaching a Tipping Point

Another decision may have had a solid rationale, but the particulars raised some eyebrows. In January 2005 under Claflin's leadership, 3Com acquired an intrusion prevention company called Tipping Point.[117] 3Com paid $442 million for the company. So why did 3Com acquire a cybersecurity firm? According to

117. Tipping Point was led, at the time, by James Hamilton, a former 3Com reseller sales representative, who had worked for Rich Redelfs at 3Com in 1990. James came back to the mothership, 14 years later! In some views, James was well liked by everyone, and was referred to as "James Bond" by his friends. Perhaps too well liked—he was also known for lavish parties for his team, which may have led to Ron Sege's replacing him with Alan Kessler later.

Anik Bose, after H3C began to take off, H3C being the first pillar of Claflin's vision, this acquisition made sense in the context of his second pillar, which basically was, "Let's get into a leading edge security product that runs inline, inside the network, so that we can use it to rebuild enterprise customer relationships and pull through the sale of H3C switching and routing gear, and help rebuild the company."

3Com also thought with Tipping Point it had an entrée into stellar world class customers and CIOs they didn't know already. In retrospect, it turned out that the people buying security were not the same people that were buying networking. The Tipping Point sellers though, were not going to point that out prior to the sale to 3Com!

Performance Re-review

As mentioned before this odyssey, there were some pre-ordained decisions that affected 3Com during Claflin's tenure. 3Com's sales DNA was slanted towards the indirect sales channels. In the early 1990s, 3Com was late to the party with its core internetworking products: NetBuilder II, and later CoreBuilder II (both late responses to Cisco). There was the fateful USR transaction, and the unfortuitous exit from enterprise products later in 2000. By the time Claflin became CEO in early 2001, he was confronted shortly by an economic recession. Nevertheless, it's viewed that under Claflin's leadership, 3Com was unable to capitalize on what was left of it, and that 3Com squandered great technology, technologists, a seasoned executive team, a customer base, and more.

Here is a short recap of what alumni shared about Claflin. At the end, Claflin will have the last word:

Recruitment and Retainment: This criticism was top of the list for many. No doubt it's hard to hire great talent to a sinking ship, but it's still the leader's role and skill to recruit well, and retain key executives, especially those with networking technology skills. Virtually all the executives left after his coming on board as CEO, within a year or two, taking with them an average tenure of seven plus years in the business.

Culture: 3Com's culture, descended from the HP alumni that arrived at 3Com, encouraged a culture that was entrepreneurial, principles-based, and technology-oriented. Those he recruited had more of an "East Coast mentality." It's a generalization with some truth behind it: East Coast companies tended to be more contentious and focused on internal politics and turf battles. West Coast companies' staffs were more free spirited and comfortable with startups needing a helping hand. And 3Com's culture was rooted in the oft-revered "HP Way." "Managing by Walking Around" or MBWA, a maxim handed down by HP, was part of 3Com's ethos. "Collaboration was missing," according to one executive.

Fit: Claflin's positive traits—smart, charismatic to some, humorous, strong managing upward—were not as useful as skills needed for managing downwards and getting through such a difficult time of restructuring and layoffs.

Teamwork: Not listening to others, along with not being trusted by others, along with an expectation of loyalty, all made for poor teamwork.

Innovation: As noted earlier, 3Com alumni felt Claflin did not foster technological innovation nor have an instinct for development and delivery on new technologies. Claflin disagrees with this statement, and pointed out numerous product introductions, based on his supporting his divisions to take innovative risks that didn't always turn out well (the DSL router with the famous 'shark' design that resulted in packaging problems, the IntelliJack, that was met with resistance by the building industry, Audrey, the ahead-of-its-time iPad-like Internet appliance, etc.). But at the end of the day, Claflin's tenure was about survival, with H3C the important and critical silver or at least perhaps chrome lining to an otherwise dismal and controversial period.

Bruce's summarization closes this odyssey: "I was dealt a tough hand, I played the hand creatively and aggressively, but did not accomplish all I had hoped or expected. As for how I approached the job from a management standpoint, I was intent first on saving the company, then focusing it on more attractive opportunities, and last, changing a culture that had gone off the rails. Progress on all fronts—but nowhere near enough."

"If you don't retain executive management, you don't retain the rank and file."[118]

Charles Giancarlo, SVP of Cisco, and Kalpana Co-Founder

Chapter 19: Huawei Wowie

Amid the cost-cutting and fire sale tactics, however, a new strategy was emerging. The endgame for 3Com would eventually be what became known as the "China Out" strategy. In time, the Chinese company Huawei would become a crucial manufacturing partner, a conduit to vast Asian markets, and an ally to help 3Com stabilize its finances.

Huawei, in 2003, was a company based in Shenzhen, China. It initially sold phone switches, but expanded into telecommunications networks, services, and equipment for enterprises inside and outside of China.

The spark for the H3C partnership with Huawei began without Huawei's involvement. As Bruce Claflin tells it, it began when Anik Bose introduced him

118. *Inside Cisco.* Pg. 204.

to the CEO of UTStarcom late one Friday night. Bruce recalled that he "had no idea who the hell UTStarcom was—it sounded like the Jefferson Starship to me." But after Anik pressed the issue, Bruce met with them. "I had low expectations, but it really was the beginning of a massive change in my mind."

He wound up spending a week with the CEO, visiting the company's operations in Taiwan, China, and in Hong Kong, meeting their customers, suppliers, and government officials. UTStarcom had large operations in China and sold their products primarily to telcos. Having lived and worked in China during his IBM career, Claflin understood the enormous potential for the China market.

"China was an awful place to do business. But UTStarcom had a fascinating business model," Bruce said. "And I came away from that trip blown away about the potential to do R&D in China." Claflin had lived in Asia with IBM China reporting to him from 1989 to 1992, when doing business with China was extremely difficult. He saw how much things had changed and recognized the new attractiveness of the market.

In fact, 3Com considered merging with UTStarcom, but Masayoshi Son (the founder of Softbank in Japan and a major investor in UTStarcom) weighed in against it; and Bruce and Anik were also not one hundred percent, so the idea was dropped. Nevertheless, Bruce was convinced 3Com needed to have a China strategy.

Claflin reflected on the value Huawei brought to the table:

> Huawei brought us an instant product line, it took us from the lowest- to the highest-end switches, and from the lowest to highest routers. That would have taken us years to do on our own. It also gave us a very large, but relatively low-cost and efficient R&D team. And finally, it got us a strong position in the China market—we went from no position to number two very quickly. And that turned out to be the most important factor in the near term.

He recounted his pitch to Huawei:

> We want to reenter the enterprise. We have a tarnished but well-known brand in the enterprise, we have channels, we've got distribution all over the world, and we can instantly help you expand and get over the brand issue for Huawei by using 3Com. Conversely, we have no real presence in China and you do. Using the Huawei brand can help us penetrate there. And last, we want to use your R&D assets as the core development site for us not only because they're skilled and hardworking, but also because they're so affordable.

> At that time, it was about 75 percent less cost per engineer. For all those reasons, we decided we wanted to do this venture. The timing was perfect, and we began the in-depth and intense negotiations with Huawei, which ultimately led to Huawei 3Com.

3Com contributed $160 million to the venture; Huawei put in 1,800 engineers. Soon, they were off to the races together to rebuild the enterprise businesses that had earlier been jettisoned to Extreme and Motorola in 2000.

The China Syndromes

While Claflin may have recognized the value of a China strategy when the prospect first presented itself with UTStarcom, pulling together all of the pieces for the Huawei venture was a complex and risky undertaking. As if the business risks and tradeoffs weren't already complicated enough, there would be some nefarious cloak-and-dagger goings on to contend with involving John Chambers.

At a time when the U.S. and EMEA markets were tumbling, China represented a green field. However, a U.S./China venture was unorthodox, and would entail risks on both sides. The liaisons from Huawei had invested in their relationships with 3Com—especially with Anik. Over time, they would grow to trust him after working with him over long hours for three consecutive months. Anik appreciated that Claflin acknowledged the vision behind the China strategy and would tell Claflin, who had previously lived in China and Japan, "Bruce, you've got one brown eye and one blue eye." Claflin understood enough about Asia to see the potential and the need for a local partner and gave Anik the money and support that he needed to proceed. Not all CEOs would have taken this risk based on the unusual nature of the agreement and the uncertainties involved.

The strategy was to focus exclusively on China, and build up a next-generation routing switching portfolio to be 70 percent cheaper than Cisco. After gaining market share in China, the product would be sold by 3Com in the U.S. Before deciding on Huawei as the best partner for 3Com, Anik had meetings and calls with a number of other companies, including Legend and ZTE. But the opportunity—and the chemistry—with Huawei stood out for Anik.

At Huawei, he presented 3Com's ideas to the Head of Corporate Development, Guo Ping, who is now Huawei's Rotating CEO and Deputy Chairman of the board. Anik later had a four-hour meeting with Huawei's founder Ren Zhengfei, who was then CEO and had been an engineer in the People's Liberation Army. He told Anik about his vision, which included the enterprise IT industry coming to China within the next year. Together they would offer an alternative to paying 80 percent gross margins to companies like Cisco. Ren told him China had a big market, and what Huawei was doing in the carrier market, 3Com and Huawei could jointly do in the enterprise market.

Anik was blown away. "He basically said, 'Hey, you work with us, we'll make it successful.' I called Bruce that evening and I said, 'I think we have found a partner.'"

There were many details to pin down. How to structure the company in a way that would make sense to both an American and a Chinese company, where to locate the headquarters for the venture, how to position the existing brands, where to locate R&D, what 3Com should do with existing facilities, like the engineering team in Hemel Hempstead. It was a complex deal.

When the China strategy was first presented to the 3Com leadership team, corporate counsel Mark Michael was dubious. He felt that their Chinese counterparts were not trustworthy. Mark even encouraged Anik to take and use machines generating background white noise on his trips to China, because he assumed every conversation might be bugged.

(Mark was hardly alone in his suspicions about Huawei. The company has since been barred from certain markets, including bidding on U.S. government contracts, discussed further below. On the other hand, in 2014, Huawei became the first Chinese company to land on Interbrand's list of the hundred best global brands.)

In addition to internal opposition to the venture, there was also external pressure hard at work to derail the partnership. While Anik was still in the thick of negotiations with Huawei in 2001, the two companies made plans to finalize and announce the deal on February 18. In an attempt to blow up the transaction, Cisco's John Chambers personally flew to Shenzhen, met with Huawei founder Ren Zhengfei, and told him not to do this transaction with 3Com, but to sell the company to him instead for $200 million, and to let Cisco absorb Huawei's enterprise business. Fortunately for 3Com, Ren Zhengfei declined the offer. Chambers flew back and promptly filed an intellectual property lawsuit against Huawei. Presumably, the real target of the suit wasn't Huawei; rather, he most likely hoped the lawsuit would be enough to deter 3Com from going through with the deal.

Unbeknownst to Chambers, Anik and his team had done extensive code analysis of the various components of the millions of lines of software used in Huawei's products. There were some instances where Huawei's engineering had "borrowed" elements from Cisco, such as the command line interface, but they were planning to remove that and replace it with 3Com's command line interface, as part of the overall joint venture agreement. Huawei also used one routing protocol similar to Cisco's, which would also have to be replaced.

Huawei was anxious about the saber-rattling from Cisco, but Anik knew 3Com had a great patent portfolio. Bruce Claflin had a call with John Chambers that played out along these lines: "John, we respect IP, and if you move forward, you will find there are no IP code issues, and we have many more patents than you, which we will use to defend against your claim." From the call, Claflin also surmised that John Chambers didn't necessarily have any great love and admiration for Huawei, not fully trusting them, which was interesting, since John was prepared to buy them out to stop 3Com earlier.

Cisco came after Huawei aggressively, first in John's private overtures and then in litigation filed in court in Marshall, Texas, which alleged infringement by Huawei of literally every relevant piece of technology that Cisco had protected. 3Com brought in a Stanford professor as an expert witness to analyze the entire code base and routing protocols. He did a complete comparison between Cisco's and Huawei's software, and demonstrated to the court there was no violation of IP rights. However, he did confirm the routing code and the command line interface would have to change.

3Com settled the case on H3C's behalf successfully. It was one of Mark Michael's last successful projects before leaving the company.

Despite succeeding in damaging Huawei's reputation for some time over the bad press involved with litigation, Cisco paid a big price of its own—losing favor in China. Their market share dropped from 40 percent to eight percent, as the Chinese market shifted toward the local favorite. In the end, Cisco didn't win either the legal battle or the war over China.

One reason Cisco had been so alarmed by Huawei's infringement was that Cisco's entire source code had been separately stolen by Russian hackers.[119] And notably, several million lines of Cisco code (a complete copy of the TCP/IP module) were contained in the Huawei software operating system, although Huawei had independently developed at least some of the other sections. Just as 3Com had done with the Palm IP theft, Cisco had never revealed this publicly, but they went on the warpath against Huawei when it appeared to relate to their IP loss.

As the venture in China grew, the last of 3Com's manufacturing being carried out in Ireland was no longer needed. The 3Com plant in Blanchardstown, Ireland, which had opened in 1991, and at its peak had employed over 1,000 people, was shut down in September of 2003, culminating in a loss of 640 jobs.[120] As H3C began to scale, the engineers in Hemel Hempstead, working on layer 2 switching and the original stackable hub genesis were no longer

---

119. Leyden, John. "Cisco source code theft part of mega-hack". The Register. May 2005.
120. "3Com manufacturing wing to close with loss of 640 jobs". The Irish Times. September 2003.

needed. It came down to cost; the engineers in China were 20 to 30 percent cheaper than the cost of an engineer in the UK.

Hemel Hempstead took another devastating hit in what was the biggest peace-time blaze accident in the UK, when a fuel depot next door went up in flames on December 11, 2005, and wasn't put out for four days. Many of the corporate offices in the area (including 3Com's) were damaged. Explosions were heard 125 miles away, and the event registered as a 2.4 on the Richter scale.

By 2005, the annual revenues from the enterprise business for 3Com were $500 million out of $651 million in revenues (the lowest overall revenues since 3Com's 1993 fiscal year, twelve years prior) and grew to be $577 million out of $795 million in FY 2006, Claflin's last year, with Tipping Point and VoIP contributing the remainder of the revenue.

"Bruce had an enormous impact on my life. Why? Because, at the time he joined, I had been president of Palm. And what I used to do is fly around, go places, eat room service, go to bed, rinse, and repeat. I had reached a point where I would sit in First Class and still be close to having to ask for a seat belt extension. I was 341 lbs. When the news came that this guy from IBM was coming, I felt that I would have to make some personal changes in my life. Started exercising, got on the discipline, got down to 192-200 lbs. And I'm at about 200 lbs. today. So, he probably saved my life or dramatically improved my quality of life without even knowing he did it."

Alan Kessler, SVP 3Com, President, Palm

Chapter 20: Endgame

Claflin left in January 2006. Scott Murray, who had worked as a Bain Capital-backed portfolio company CEO among other roles, stepped in to take his place. Scott actually wanted to sell 3Com's 51 percent interest, instead of the reverse occurring with 3Com buying out Huawei's 49 percent interest in the H3C venture. He had serious concerns pertaining to H3C's business ethics, along with potential cyber security risks.[121] Scott's plan was to use the funds to make more relevant or appropriate acquisitions in the U.S., in particular investing in security to build on Tipping Point's success. As Scott noted:

> 3Com was a really good platform, with some good bones to it, but when you sifted through it all, you had a security business, an SMB business, and a Chinese business. The security business is where I would have put my money on, build up a presence. The board and Eric doubled down on the switch and router business, and I decided to not participate.

Eric's scenario prevailed and Scott departed seven months later in August of 2006. Edgar Masri returned to 3Com as CEO in August of 2006, stepping in to execute the buyout of H3C's interest, and build up the momentum back in the U.S. This set the stage for 3Com to fully return to the battlefield against the likes of Cisco and Nortel in the enterprise.

During Scott's tenure, Huawei had hired investment bankers to both respond to and solicit interest in the purchase of H3C. Sensing the high desire in investing in Chinese high tech companies at potentially rich valuations and noting the desire of H3C's management to operate independently from 3Com, Huawei focused its efforts on selling its share and convincing 3Com to sell its own share to a private equity firm. Anik Bose, Neal Goldman (3Com's legal counsel), and Edgar Masri, who stepped in as CEO after Murray left, worked tirelessly to turn the tide against this outcome, aiming to bring H3C fully into the 3Com fold. This was the most significant event during Masri's tenure.

Masri is outgoing, curious, has high energy intellectually, and is a talkative extrovert. In conversations, he sometimes joked about how the street skills he gained growing up in Lebanon later helped in dealing with the 3Com politics as CEO. This type of high energy personality could sometimes be controversial in an engineering-centric company. While attending 3Com's "Billion Dollar Manager" management training class in the mid-1990s, while a top engineer himself, Masri had quickly offended all the "introvert" engineering managers with his high energy, extroverted personality.

Masri started his career at Bridge Communications, where he learned about terminal servers and routing, before attending Stanford Business School. Eric later talked Masri into returning to Bridge, and Masri returned the favor with his loyalty to 3Com. At 3Com, Masri started out working on software,

---

121. McKinnon, John D."U.S. Weighs Curbs on Chinese Telecom Firms Over National Security Concerns". The Wall Street Journal. May 2018.

including the OS/2 and LanManager products. At the time, Eric was the general manager and Alan Kessler was Director of Product Management.

When he was product manager for the LinkSwitch product, Masri built up a good reputation with colleagues, including Andy Verhalen and Eric Benhamou. LinkSwitch was a good product for both workgroup and desktop environments, whereas Grand Junction, which Andy had been investing in as a venture capitalist, was solely desktop-oriented. (Larry Birenbaum later said it was a dark day for Cisco and Grand Junction when 3Com launched its LinkSwitch product.) LinkSwitch had great engineering resources from a laid off DEC team in Ireland, which gained them access to key ASIC technology and helped 3Com "cross the desert" in 1992 and 1993.

For Edgar Masri, landing the 3Com CEO role in 2006 involved some old-fashioned serendipity. Eric had worked with Masri previously and liked him. Around the same time Scott Murray was being hired, Masri reached out to Eric for the first time. When things didn't work out for Scott, Eric and Masri joined forces. Masri had been studying Mandarin on his own, due to his personal interest in the country and the movement he saw in the VC community to conduct business there. Masri already spoke five languages; why not one more?

Masri felt that Bruce had done a good job in establishing the joint venture with H3C and that it had been a good attempt to resurrect 3Com as a player in the enterprise. In 2006, after taking over the CEO helm, Masri successfully bought out Huawei's remaining 49 percent share in H3C for $882 million.[122] As noted earlier, this was no small feat, and an important step and big win for 3Com and Masri, since they beat out competing private equity firms, such as Bain Capital, Silver Lake Partners, and Huawei itself. Masri noted the key reason Huawei ultimately allowed 3Com to be the buyer was that they were a large, well-respected, intellectual property company and would carry clout in case of IP disputes with large companies. Other buyers did not have this stature in technology.

The Bain of Edgar Masri's Existence

During Edgar's Masri's tenure as CEO, he talked with Harvard Professor Joe Lassiter, an old colleague of both 3Com founders Howard Charney and Bob Metcalfe (Joe had known Howard and Bob in college). Masri and Joe discussed the emerging opportunities for 3Com. Joe, who served as faculty chair for Harvard Innovation Lab, shared with Masri his own thoughts about the H3C buyout and the Bain transaction.

---

122. Hochmuth, Phil. "3Com buys out Huawei joint venture for $882 million". Network World. November 2006.

Around 2006, Joe began working with 3Com after reconnecting with Edgar Masri through a student of his at the Harvard Business School. At this time, he remembers, "3Com was struggling as a company. Everything they tried had backfired." The acquisitions of USR (except for some cash generated from the Palm spinout) and Tipping Point were not particularly helpful to 3Com, overall. In Joe's opinion, these acquisitions signaled to the marketplace that 3Com was going to focus on small to medium-sized enterprises. This raised an unintended red flag warning to big companies—or those aspiring to be big—to avoid doing business with 3Com, even for products well-suited to their needs. Instead, they sought out Cisco solutions. Unsurprisingly, 3Com also saw a flood of talent leaving its ranks as many staff moved on to more promising ventures.

Masri was debating two strategies to address the company's shortcomings and confront what had become an uncertain future. The first idea was to break 3Com into its various components, and sell off individual parts. Masri's other strategy was to go big, embrace a globalization strategy, and expand the company in emerging Asian markets. Joe recalls Masri's perspective at the time, "Edgar had a pretty straightforward view. He thought he could hold his own against Cisco in China, if not outperform them, by using the H3C salesforce." The crux of Edgar's new strategy was to source low cost—but competent—engineering and compete in a market where Cisco "couldn't beat the crap out of him."

The goal would be to build 3Com up again and counterattack in emerging markets, if not in the U.S. and Europe. The problem, Joe noted, was that "no one really knew how to merge the cultures of a Chinese company and an American firm successfully."

With the advice from colleagues in the venture business who also understood the private equity business, Masri decided that it made sense to take 3Com private. Over the course of a few years, the company could be cleaned up and reinvigorated, and then made public again or sold to a different buyer. At the time, Joe believed this could be a successful strategy, but it would require Masri to successfully convince 3Com's board, customers, investors, employees, and ultimately American regulators, of the plan's merit.

As Masri shopped around for a private equity partner, he found that some firms favored the plan to break up and sell off 3Com's various components, while others favored the plan to "go big." Bain Capital Partners favored the plan to "go big," and was eventually selected as the right firm to buy 3Com out from under its public shareholders. In 2007, 3Com agreed to a $2.2 billion buyout by Bain, which would also give Huawei Technologies a minority stake in the restructured company. This proved to be a very complicated transaction—one which would ultimately fail, given the nature of working with governments and handling the integration of the two organizations.

Masri had his own specific concerns. He was worried about the possibility that the Chinese government would agree to the transaction, let everything become integrated, and then take the company back by forcing Huawei to reacquire H3C and appropriate 3Com intellectual property. Joe agreed there was "every reason to believe that the Chinese government might do that."

But Masri also knew the timing was right for Huawei to participate in this transaction. They were competing against Cisco and their network business was imploding due to inefficiencies in the equipment being pushed by the dominant European and American vendors. Masri also believed that there was a "mutually assured destruction" safeguard that would dissuade Huawei from screwing 3Com. Doing so would fundamentally affect their ability to execute deals with other Western companies, and the Chinese government would also be concerned about hurting other Chinese companies pursuing expansion into global markets.

While the Bain deal was still on the table, mutual skepticism sizzled between the Chinese and American parties. The Americans were concerned that bringing a Chinese company "inside the tent" would invite trouble, especially when involving products with intrinsic security considerations. The Chinese were suspicious their American counterparts weren't going to treat them as full partners.

And security considerations aside, in the American business culture of the day—in particular, the venture capital community—the Chinese business community was viewed as a risky bet. Accordingly, there was skepticism about whether the H3C venture could ever gain a global presence. This also complicated the sale to Bain (and, according to Masri, also made the integration of the 3Com and Huawei sides of the joint venture very difficult).

These factors obviously strained the Bain deal, but consensus identifies the main reason it went off the rails was, aside from generally poor execution of the deal, anxiety surrounding the requirements of the Committee on Foreign Investment in the U.S. (CFIUS). Chaired by the U.S. Secretary of the Treasury, CFIUS includes representatives from the Defense, State, Commerce, and (today) Homeland Security Departments.

CFIUS issues might have been resolved faster if Bain and 3Com had simply asked CFIUS what they required to green-light the deal. (The answer might have been the deal was a non-starter, but they would at least have known much sooner.) Instead, they wasted money and cycles by offering up various olive branches during the process, such as reducing the equity stake that H3C would have in the new entity to under 10 percent. In reality, however, CFIUS would have none of it. Even a 1 percent equity stake would not have placated them, according to Masri.

CFIUS representatives clung to deep-seated fears about digital infrastructure products made in China being sold into the U.S. Ironically, the Bain buy-out would have actually given 3Com more control over the Chinese venture; the proposed deal would have left Huawei with only a 10 percent stake, far less than the 51 percent or 49 percent ownership it had during different stages of the joint venture with 3Com. And during those years, Huawei already had access to all of the technology that wound up in 3Com products anyway; so, blocking the deal with Bain did nothing to shield that technology from a company that CFIUS feared was too close to the Chinese government. There were also some shenanigans regarding the bonus that an H3C executive wanted, which may have disrupted the buyout activities. But this wasn't the deal-killing issue that CFIUS proved to be.

Joe Lassiter felt that the U.S. government had been worried about the ability of a Chinese company to insert back doors into equipment that could access and alter Western networks. In an article in USA Today, Donna Borak wrote that lawmakers and Bush administration officials expressed concerns "that sensitive military technology could be transferred to China through the 16.5 percent 3Com stake that would be held by Huawei."[123] Bain had proposed divesting 3Com's Tipping Point subsidiary—on the assumption that their network security software might be a point of particular concern—but it was not enough to assuage the concerns of the committee. 3Com eventually withdrew its voluntary CFIUS application and was unable to further pursue the transaction.

At the time of the failed Bain transaction in 2007, there was Cisco and the 'seven dwarves'[124] competitors—Juniper Networks, F5 Networks, Foundry, Alcatel, Extreme Networks, 3Com, and Nortel. Only three dwarves remain standing today:

Those still with us:

- Juniper, with its ground-breaking routers, has $5 billion in annual sales for 2017 and 9,500 employees.

---

123. Borak, Donna. "Bain-Huawei offer for 3Com on the rocks". USA Today. February 20, 2008.
124. For history buffs, the seven dwarf reference was actually used earlier with a group of mainframe computer competitors to IBM in the 1970s—Burroughs, UNIVAC, NCR, Control Data Corporation, and Honeywell. These along with RCA and General Electric, were called IBM and the seven dwarfs.

- F5, focused on services and software, continues today with 4,300 employees and $2.1 billion in annual sales.
- Extreme, a beneficiary of 3Com's exit from enterprise, today is an enterprise networking company with 3,000 employees and annual sales of $600 million.

The departed, in addition to 3Com:

- Foundry, a switch and router company, at one time valued at over $9 billion, was sold in 2008 to storage networking company Brocade Communications Systems for $3 billion.
- Alcatel merged with Lucent in 2006, and that company combined with Nokia in 2016.
- Nortel, a giant in telecommunications and networking, with 95,000 employees at its peak, filed for bankruptcy in 2009.

As an interesting footnote to this competitive landscape, Leon Woo, (architect of Synernetics' products in the early 1990s that led to 3Com's successful Switching Division, later landing at Enterasys), worked with Masri later in 2006 on a novel idea to take on Cisco: a merger of several networking companies. Woo explained:

> This is probably not known, but there were discussions with four networking leaders in discussions to merge—Enterasys (from the breakup of Cabletron), Juniper, HP Networking, and 3Com, to create a credible number two to Cisco. Cisco has already won—we had no thoughts of creating a new number one in networking. Cisco "out acquired" the rest of us largely due to efforts of Mike (Michelangelo) Volpi, the business development genius under Morgridge and Chambers. We got close to HP but were unable to finalize it.

A Slow Boat to China

Masri ultimately left 3Com as a result of the impossible realities of managing China from the U.S., coupled with the failure of the Bain transaction. Without the restructuring that would have happened after the Bain deal, the Chinese venture—the basket into which 3Com had carefully and intentionally placed nearly all of its eggs—was stuck in neutral.

A major source of friction that had hindered the H3C relationship was the hiring of Bob Mao as Vice President of Business Development under Masri. This wasn't due to any shortcomings on Bob Mao's part—indeed, he would ultimately become an effective leader for H3C (one executive described Mao as "an emperor but also a wiseman". But his role hadn't been positioned correctly in the eyes of the Chinese partners, who viewed it as a serious threat: this Taiwanese guy gets hired and the H3C executive Shusheng Zheng (aka Dr. Zheng) is suddenly fearing his own replacement.

At the founding of the H3C venture, Dr. Shusheng Zheng was Huawei's first CTO under Ren Zhengfei. He helped develop H3C's original product—the SS7 switch. Zheng ran the Huawei Enterprise Division that eventually became the H3C joint venture. He was a very smart technologist and driven executive who helped grow H3C from nothing to 6,000 plus employees. He was not particularly polite, and he believed that the China way was the best way, which didn't mesh well with western customers. According to executives, Zheng's work was underrated at 3Com because of his resistance to taking direction. Nevertheless, he went on to found three companies in China that are rumored to be doing well.

Exacerbating the relationship issues between East and West was the false hope among the Chinese team that Edgar would be a "yes-man" and would accept any and all ideas put forth by H3C. Masri was not a "yes-man," and, as a result, the Chinese team tended to block his every move.

Bob Mao knew how to manage the issues and did a good job, according to Masri, although some have reported Mao was not very easy to get along with. He was physically present, and he spoke the language. He also knew what made them tick: bonuses and profit. And the best motivators in China could be different than those in Silicon Valley; instead of asking "How many stock options can I have?" engineers in China asked, "What kind of car do I get?"

For Masri, the experience was a source of constant frustration. He traveled, spent time in China, and spoke some Mandarin. But friction still arose, particularly when it became apparent that China also wanted to control the P&L for the entire Asia Pacific region, including South Korea, Australia, and other markets. In hindsight, Masri felt he should have handed the reins to Bob Mao sooner, which is exactly what the board did after Masri left, hiring a package of Bob Mao as CEO, while naming Ron Sege as the COO and President, running all things outside of China.

There were political challenges as well, with China wanting to take over more of the product, while Masri dealt with product redundancy between Marlborough-based products and H3C. Masri may not have been decisive enough to consolidate the legacy product in Marlborough with H3C, which is what Ron Sege later did. Border skirmishes about which countries were being sold products were constant (for example, China moving product into Eastern European countries while the American salesforce was responsible for that territory and OEM'ed their products from H3C to resell).

Masri noted:

> It didn't matter much what 3Com in the U.S. was doing at the time as long you kept building up China. They were building this great switch router, and when it was done, it really caught the attention of companies like HP and others. And the board and Eric, we all realized we needed somebody right there in China to manage the Chinese, and someone here to manage the U.S. I tell you, the solution they implemented with Ron and Bob Mao, was fine, and the best strategy.

Masri left in April 2008, after the collapse of the Bain transaction. When he joined as CEO in 2006, the company had closed with $800 million in sales, with 5,600 employees, 4,000 in H3C. For 2008, 3Com had $1.3 billion in revenue and more than 6,100 employees, with 4,700 in H3C. 3Com had done what seemed impossible in 2001 when Claflin took over as CEO.

In the 1990s, Masri was viewed as a key force behind the success of PDD in the UK. Around 3Com, he was considered very smart, hardworking, and with high integrity. Eric especially held Masri in high esteem. But as CEO, the board felt there were various business decisions that didn't go well. Besides the difficulties he had with the Chinese, a closer-to-home issue involved the idea of taking Tipping Point public, which didn't materialize. Another involved 3Com's VoIP product decisions around NBX, which Ron Sege later eliminated. But no doubt, the Bain collapse was a key tipping point. The board decided they needed to make a change, and further build out the business strategy. They selected Ron Sege as the President and COO, and named Bob Mao the new CEO to be based in China.

The Sege Segue

Ron Sege grew up in Silicon Valley where his father Thomas Sege, arrived as an immigrant from Yugoslavia. Thomas was a microwave tube engineer who eventually ended up running the well-known early Silicon Valley giant Varian Associates from 1981 to 1990. Varian was the first tenant in the Stanford Industrial Park, even before Hewlett-Packard, which was the first lease holder.[125] Ron's father retired at sixty-five, working into the late 1980s. He was a principal

---

125. Stanford Research Park was built in 1951 as Stanford Industrial Park, and Frederick Terman was credited with its founding. As the first university-owned industrial park, it contributed to the birth and rise of Silicon Valley. In 1953, Varian Associates became the first tenant. It covers 700 acres and is still home to Hewlett-Packard, various legal offices, and previously housed Facebook. Fred Terman was dean of Stanford's School of Engineering. He helped expand the sciences, statistics, and engineering departments, and secured grant money from the Department of Defense, elevating Stanford's stature, while helping to fuel Silicon Valley's growth.

developer of the klystron tube, used in radar and TV transmission. It's easy to guess where Sege inherited his stamina and drive to achieve. One of Ron's early pivotal experiences was his ascension to the leadership of the nation's Junior Achievement organization as a senior in high school. In this role, he oversaw 6,500 students at the entrepreneurial group's annual conference.

After college Sege worked with economist (and Fed Chairman under Carter and Reagan) Paul Volcker, helping to evaluate wage, price, and productivity information using the miracle of Fortran 4. From there Sege joined ROLM Mil-Spec Computers as a summer intern, and Bob Finocchio later hired him into ROLM's Telecom business. Finocchio suggested to Sege that "Harvard MBAs can either go into marketing, finance, or get a real job and go fix the customer service organization." Sege took the real job. He recalls jumping in head first, figuring out how to relate to and get along with the telephone installers hired from Pac Bell to install their PBXs.

Ron stayed with the company after it was purchased by IBM, but left when it was reconfigured into a joint venture with Siemens. In 1988, Bob Finocchio came into 3Com, hiring a massive sales and service organization, including many hailing from ROLM. After landing at 3Com, Ron Sege ran the customer support organization, and after the company bought BICC, he ran PDD in Hemel Hempstead. He later took on responsibility for Chipcom, and the role eventually moved him to Boston. When Finocchio left 3Com, Sege took over the Network Systems Operations Group, while Doug Spreng ran the adapter business. Ron is an INTJ on the Myers-Briggs scale and his leadership style is friendly, approachable, decisive, participative, and consensus-oriented.

Ron quit 3Com earlier in 1998, when it was clear Bruce Claflin wasn't asking for his help. When the Bain transaction under Edgar Masri fell apart at the last minute, there was a mad scramble by the board to figure out how to proceed; the biggest challenge was to figure out how to manage China. Sege initially joined the board in 2007 after Claflin's departure while Masri was CEO, and became the company's President and COO in 2008 after Masri left the company. Bob Mao was named CEO.

Initially, expectations were unclear, and communications were distorted. The H3C team in China was puzzled and frustrated as to why the 3Com in the U.S. was running parallel operations, rather than simply selling the H3C products. The dual leadership idea was invoked, giving Bob Mao full responsibility for the H3C venture, as well as the title of CEO of 3Com.

Ron's strategic viewpoint was actually more closely aligned with that of the team in China. The operations in Marlborough, MA only existed for legacy reasons. Sege solved it by moving all the R&D to China, beginning what became the now, at least internally, famous "China Out" strategy. This also removed any border skirmishes with the Chinese selling into U.S.-based sales rep's territories, as H3C would now have all revenue credit from their manufacturing site.

Masri had earlier encountered the Chinese culture, arrogance, and business practices, and now it was Ron's turn. Mao did not want to share responsibility with Ron, and also expected obedience. Earlier, Sege had outlined the roles and responsibilities between the COO (Ron) and the CEO (Bob). Mao felt that Sege was taking liberties trying to tell him what his job responsibilities were. Ron's tight relationship with Eric and the board may have also fueled Mao's resistance to suggestions from Ron.

One CEO action that frustrated Sege was Mao's letting go of Eileen Nelson, a strong and competent HR executive. Mao wanted to take over HR, told Sege that HR would report to him, and Eileen was let go. The Board of Directors' Compensation Committee challenged his call but to no avail. Per Ron Sege:

> Mao had no interest in the kind of HR that we do here in the West. HR in China is really about grooming loyal talent, and putting people in boxes to achieve practical objectives, it's nothing about gender equality, and the laws we've got to enforce sensitivity training, they could give a fuck about that.

No question—Mao had strong control over many aspects of the company. And H3C and Bob Mao were not without controversy. One controversy related to H3C finding creative ways to take sales away from 3Com's own Tipping Point sales team, including sales in territories covered by U.S. or Europe sales reps, for the benefit of the H3C team, much to the consternation of the Tipping Point team.

The China Out strategy was aligned with Clayton Christensen's notion of disruptive innovations. This kind of approach often begins with low-end disruptions. For example, a company trying to break into a market might find early success with products at a low price. Such products often emerge in markets like India and China where consumers may be more accepting of relatively low performance if the price is also low enough. Over time, the products improve, the technology matures, and the product is commoditized—and the countries and companies that leapt in at an early phase become the winners in the developing world.

This approach can enable new players with new products to break into new markets. For example, when the personal computer disrupted the existing market for high-end computers, the doors swung wide open for disruptive new entrants. These disruptions often follow a typical pattern. First, costs are driven down dramatically, by attacking competitors' higher cost structures and business bloat.

Following this, innovation tends to occur in the safer home markets. Next, a more modern design platform is developed that can deliver faster cycle times for new product iterations. And finally, a new caliber of service that disrupts your competitors' customer interaction model has been delivered.[126] Companies that have successfully adopted versions of this approach include LG, Samsung, and Toyota. And Huawei and 3Com saw eye-to-eye on this vision.

Ron added further, "Companies that try to change or go against their DNA fail. Ours was high volume, channel, vertical integration, while Cisco's DNA was customer intimacy." Sege also referenced an analogy from Michael Porter's theories in the '80s that had defined the three strategy types—cost leadership (think Walmart, McDonalds), differentiation (Gucci, Apple), and market segmentation or focus (blend of the first two, Southwest Airlines). Ron's taking over while H3C was taking off showed the power of combining cost leadership from their position in China, and their use of focus, targeting the customers wishing to save money on their enterprise networking budgets.

Ron also learned a few important lessons specifically about conducting business in China:

- In the West, people do a deal, and then get to know each other. In China, people get to know each other, then do the deal (the word "guanxi"—the relationships individuals cultivate with other individuals—is an important one to know for Americans doing business in China).
- China has been around for a long time, has had great success, and is a land filled with great entrepreneurs.
- Saving face matters in China.

A Forbes article about 3Com's China Out strategy got a fair bit of attention. It presented that the way the company was disrupting the networking industry was similar to how the auto industry in Detroit was facing disruption from unexpected competitors. As the hope of finding a final exit strategy for 3Com started percolating in earnest, this kind of external validation of the strategy certainly didn't hurt, and according to Ron Sege, gained them some notoriety.

But according to Ron, it was a serendipitous accident that led to HP buying 3Com. Sege tells the story:

> At Interop in 2009, as part of this China Out strategy, 3Com had introduced the 12500 data center switch. A guy named Marius Hass that ran business development at HP [and is now president of Dell]

126. "Disruptive Innovations". Christensen Institute.

was going by the booth—so I grabbed him and shared this new product with him.

A week later Dave Donatelli asked me to come by after his meetings at 7:30 p.m. in the evening. Dave had come from EMC to run systems for HP and they were planning to buy Brocade. But Donatelli had looked at it—it was 80 percent an InfiniBand company [making ultra-high-end networking products, often used with supercomputers], and they wanted a networking company. Donatelli told Marius this the day after Marius had seen the 12500 switch. Marius said, "Well, Dave, there is this other company that would give us a switch and low-cost access." Dave's view on taking on Cisco was focused on cost—HP's buying power, manufacturing prowess, 3Com's low-cost R&D. And with home country volume in China, with an in-country salesforce, they could beat Cisco.

To mangle a cliché, serendipity proved to be the mother of necessity. In November 2009, HP announced that they would buy 3Com for $2.7 billion in cash. For HP, the acquisition would help bulk up their networking product line and move ever closer towards being a one-stop shop for corporate customers as well as land them a 30% market share in the networking market in China. The acquisition was finalized on April 12, 2010.

Saar Gillai, The Senior VP running 3Com's worldwide products and solutions shared:

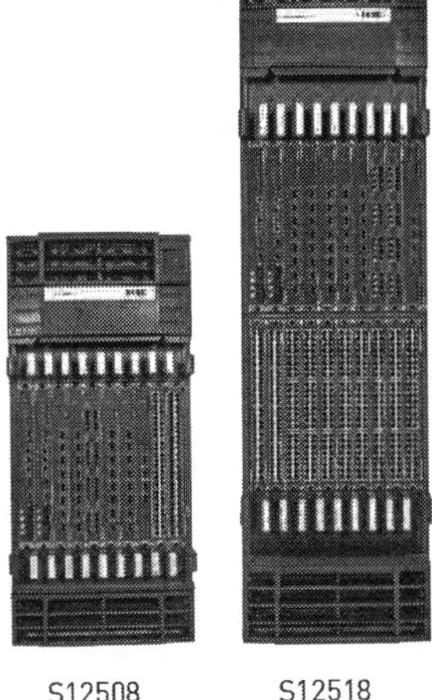

The H3C S12500 next generation large core/data center switching platform that caught the eye of HP at Interop 2009.

> In spite of Sales not wanting to spend the money, the $1+ million that we paid for Interop 2009 to introduce H3C to the West definitely worked. H3C had not really successfully entered the U.S. markets or the important Data Centers until this event. The 12500 switch was a worthy adversary also against Cisco's Nexus 7000.

Ron had already been holding acquisition discussions with IBM, Dell, and HP. And I had to frustrate the salesforce and their customers by diverting

> 12500 switch units to provide evaluation units to HP and others for their testing. At the last minute, [HP CEO] Mark Hurd held up the acquisition pricing discussions in a critical Beijing diligence meeting with Goldman Sachs and 3Com. Thankfully, Bob Mao helped smooth out the wrinkles with a handshake with Shane Robison at HP and we got it done. As all deals go, we all worked crazy hours to complete the due diligence. In summary, Sege lit the match, I helped fuel the fire, and Mao closed the door, along with the great help of many others on the diligence front."

3Com's financial trajectory may not have ended on a high note, but it did manage to deliver a palatable exist strategy.

The China Out strategy had accomplished its goal which, for better or worse, was not simply China Out, but also "3Com Out."

The After Parties

So, what is HP doing with 3Com now? According to Ron Sege, because HP's Dave Donatelli was used to selling switches and routers alongside storage systems and servers, he mashed the salesforces together, which ended up being a total integration disaster for HP. However, he left the China strategy alone. The 3Com sales guys left—their infrastructure and product management support soon were all gone. H3C did fine, and continued to grow. HP's CEO Meg Whitman called on Bob Mao to return to HP to help run greater China. Whitman came in after Mark Hurd and the short-tenured Leo Apotheker and opted to sell storage and servers branded as HP products. These were folded into H3C, and the majority of the company was then sold to Tsinghua Holdings, thereby making it (technically, at least) an indigenous Chinese company. The products were then sold back to HP on an OEM basis, which retained a 49 percent stake.

Although Cisco dominates the Ethernet switching space, Huawei and others also found success in the arena.

Today, Tsinghua continues on successfully as one of many Tsinghua University's subsidiaries, a public university in China. HP later bought Aruba, and Aruba's salesforce wound up selling what had been 3Com products. Ron Sege ran into Meg Whitman in 2017. She said 3Com and 3PAR proved to be the best deals for HP based on their internal rate of return. 3PAR. (The two "3" companies weren't related although Robert Rogers, the R in 3PAR, had worked at Bridge Communications in the 1980s.) And Huawei's success is enormous, even if controversial—as this book was being finalized, Ren Zhengfei's daughter, Meng Wanzhou, was arrested in Canada and facing extradition to the U.S. on charges of breaking American sanctions on Iran through a Huawei subsidiary called Skycom, between 2009 and 2014. She is the CFO for Huawei.

3Com's 30-Year Trajectory

3Com sales and net income, in $ millions

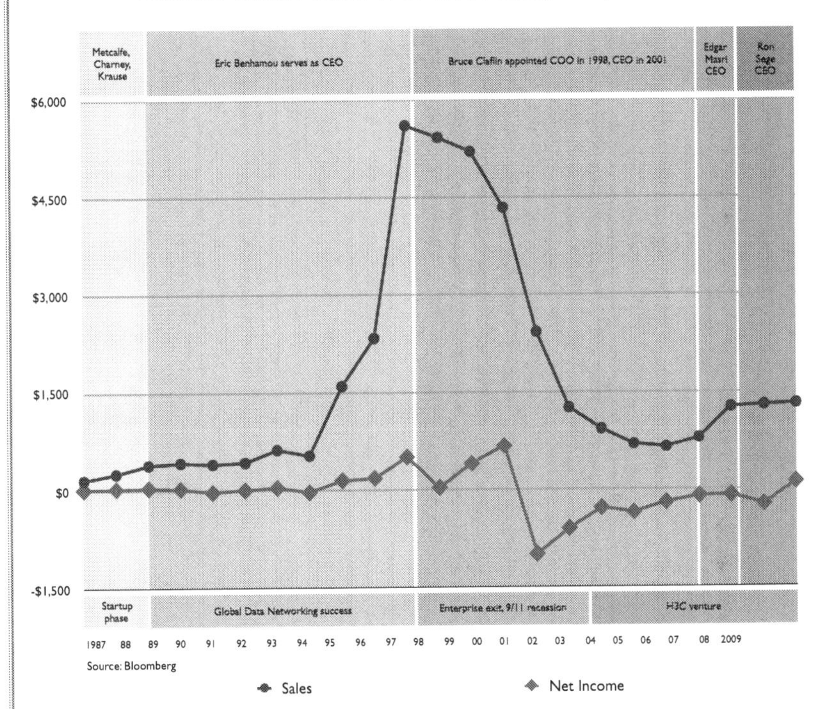

Eric's reign lasted well after his CEO role, as he oversaw the strategic direction as Chairman of the Board until the company was sold. The Metcalfe, Charney and Krause years illustrated amazing and explosive growth. Eric deserves credit for the success of the global data networking strategy, but suffered from the challenges of overreach with the USR acquisition in 1997, and the chasing after markets maturing versus expanding, (Palm's success notwithstanding). Masri and Sege did their level best dealing with China's challenges, game to carry forward Claflin's China Out Strategy, leading to the sale of 3Com to HP in 2009.

Finding Your Niche

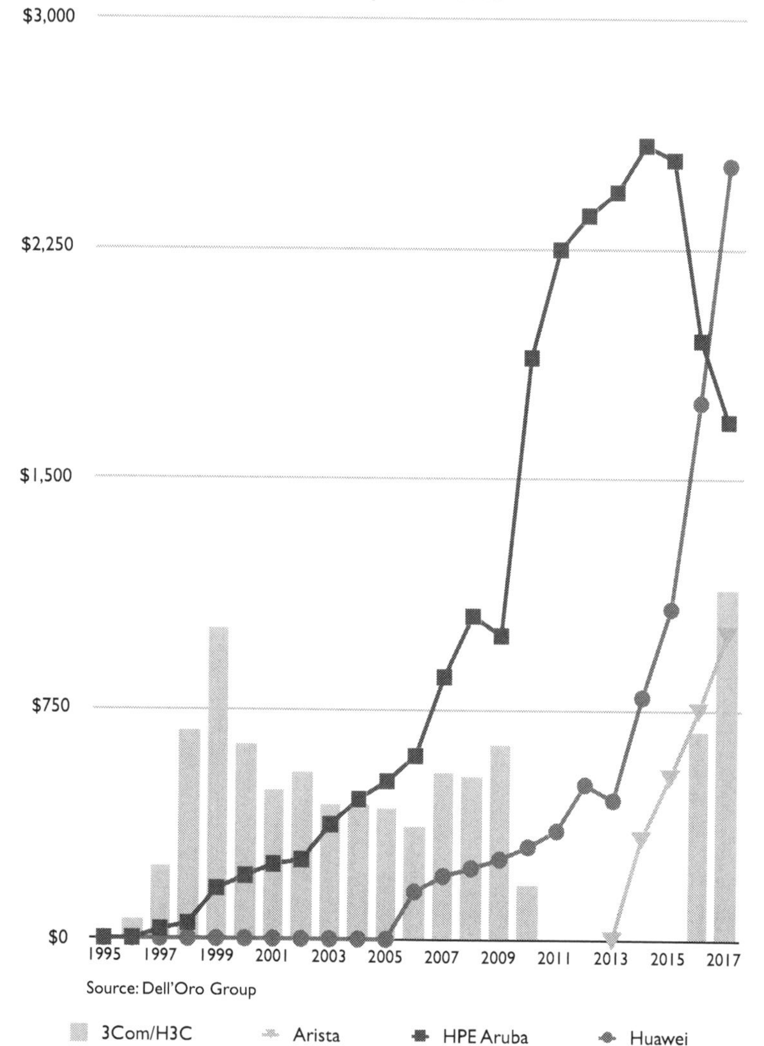

Being number one in your market is nice. But you can be successful without dominating your market. For this chart, Cisco's sales of Ethernet switches ($14 billion in 2017) have been omitted to make the also-rans more visible. And many of the also-rans (including Huawei, Arista Networks, and other companies not included in the chart, such as Juniper, Brocade Extreme, Netgear, ZTE, Dell, and D-Link) are running quite well, with sales in the multi-billions. The latest incarnations of H3C have also done well in this market. After 3Com's sale to HP, H3C settled into the HPE Aruba group. In 2015, HPE Aruba sold a 51% interest in H3C to Tsinghua Holdings in 2015, so its revenues after 2015 are not shown.

Telecommunications firm Orange also has ruled out using Huawei products in its core 5G network in France, and Germany's Deutsche Telekom says it's reviewing purchases of Huawei equipment."[127]

As for the competition? HP took aim at Cisco with the purchase of 3Com's networking products, as a counter to Cisco's move into servers, which was previously the province of HP and Dell. Oracle snapped up Sun Microsystems, folding it into their computer business. Dell bought Perot Systems, for tech consulting, which stepped on IBM's and HP's toes. And Cisco picked up Tandberg SA in a bid to counter HP's videoconferencing systems.

Meanwhile, other companies struggled. Motorola sold off its networking business. IBM sold its PC business to Lenovo Group. But HP made significant inroads into the networking market, with the catalog it picked up from 3Com and the acquisition of other companies that complemented its existing products. At the time in 2009, HP had become the distant, second largest maker of networking gear after Cisco, with 10 percent of the market (compared to Cisco's 70 percent).[128]

On July 1, 2010, 3Com and Palm had a reunion—on paper—when HP purchased Palm for $1.2 billion in cash, largely to acquire Palm's webOS operating system. But whatever HP's plans for webOS were, it got lost in the game of CEO musical chairs that was also taking place at the time. In the fall of 2011, HP shuttered the Palm operations and wrote off $885 million associated with the purchase.

3Com"s foray into the Voice over IP or VoIP space started with the NBX in 1999, a Boston company, that overlapped with a VoIP system called VCX within USR's product division. Edgar Masri as CEO reduced the investments in NBX, and continued with VCX, which survived and was moved to China under the Sege CEO era, then was terminated by HP later in 2012. Ron Sege recalled, "By the time I came back to 3Com as CEO, the NBX product was dated. The VCX was more capable, and I made the decision to move it back to China. The two products together had miniscule market share. The future was in our switches." Also, the fact 3Com was not as familiar with sales channels for this product, left them as a "me, too" vendor, while VoIP was dominated by other vendors—Cisco, ShoreTel, Avaya, Mitel, Siemens, etc. The number of VOIP users went from 150 thousand in 2003, to 1.2 million in 2004,[129] but 3Com did not take a large share of this market.

HP sold 3Com's Tipping Point business to Trend Micro for $300 million in 2015 (vs. the $430 million 3Com paid in 2004). This price delta suggests that

---

127. Riley, Charles. "Doors are slamming shut for Huawei around the world". CNN Business. December 2018.

128. Scheck, Justin. "HP to Acquire 3Com for 2.7 Billion". The Wall Street Journal. November 2009.

129. For a more complete explanation of VoIP technology, visit http://www.3comstory.com/tech.

3Com earlier paid too much for Tipping Point, perhaps using too much cash that could have been put to better use.

Cisco's capture of 3Com's top talent in the early 1990s via the Grand Junction acquisition (Charney, Verhalen, Moses, Birenbaum, Schwartz, etc.) along with its speed in adopting Grand Junction's Fast Ethernet standard into the chassis-based switches from its Crescendo acquisition, helped Cisco immensely in launching the Catalyst family of products. When ex-3Comer Larry Birenbaum left Cisco in 2004, Grand Junction's products were generating $3.5 billion a year, and commanding 65 to 70 percent of market share for desktop switches. Despite having a small staff of about 700 people within Grand Junction's business line, Cisco's phenomenal salesforce and support was able to make the product line extremely successful.

And how is Cisco doing today? Even Cisco now faces a type of innovator's dilemma, not unlike what 3Com faced in the second half of the 1990s. Howard Charney, who recently retired from Cisco from his position as SVP in the office of the CEO, noted:

> The installed base for Cisco comprises literally billions of dollars of routers and switches, and to its main tenants, this is exceedingly expensive for them to maintain on their R&D budgets.
>
> So, Cisco's customers want Cisco to make them better, make them faster, come up with and adopt new standards, say, put in a new 100 gig blade, and these are very, very expensive developments. Now, is that a bad thing? No. Except, Cisco has, depending upon which market researcher you're looking at, Cisco has somewhere between 60 or 70 percent, maybe as much as 80 percent market share, depending on which market you look at. So, it's very, very difficult to gain market share because to go from 70 points to 71 points means that you have to work super hard against your competitors who don't want to give it up. It's a really expensive game.
>
> In the meantime, markets are shifting. Because of competition like Huawei, Alcatel-Lucent, and others, it's really hard to change itself over time, without screwing the goose that laid the golden egg, and turn itself into a software and services company.

In the fall of 2017, Cisco lowered its long-term revenue growth expectations to a rate of 1 to 3 percent, citing a challenging and competitive environment that would affect its core hardware business. Cisco today continues to suffer from commoditization of its core hardware (of which switching and routing hold 61 percent market share), and services and software are not growing fast enough to offset hardware.[130] Hmm, sounds familiar!

And now we'll give Ethernet the last word. If you have made it this far, you now know that in the 1980s, Ethernet won the war over the LAN standards, defeating IBM's Token Ring, plus FDDI and ATM. Twisted-pair wiring for Ethernet helped fuel further growth. As Moore's Law showed us, machines got much faster, and more machines connected (Metcalfe's Law) in the 1990s, and Ethernet Switching exploded to meet the needs for client-server networking, allowing networks to scale in the thousands of users in the enterprise.

Gigabit Ethernet (GE) evolved to be dominant in data center and cloud server connections. By 2006, Google was deploying thousands of Ethernet switches a month. (One of Google's cluster switches provides 40 terabits per second of bandwidth, the equivalent of 40 million home Internet connections).[131] Demand on traffic exploded, based on Software as a Service applications (e.g., Salesforce) and social networking (e.g., Facebook, YouTube). WiFi now competes with Ethernet at the desktop edge of the network.

By the time 3Com was sold to HP, Gigabit Ethernet (GE) was running out of gas, and 10 Gigabit Ethernet was needed for cloud servers. Microsoft's Azure cloud exceeded a million servers, and needed to scale from GE, to 10GE, to 40 GE, with an eye to 100 GE. Facebook and Google have expressed interest in Terabit Ethernet with 200 and 400 GE standards being developed. The future is ever expanding, as new technology called Time Sensitive Networking Ethernet will be adopted for automotive, airplane, and cloud computing requirements.

Moore's Law and Metcalfe's Law are still relevant. And Gilder's Law of infinite, unmanaged bandwidth marches on.

130. "Cisco's Services and Software Business Isn't Quite There Yet". Seeking Alpha. September 2017.
131. Cade, Metz. "Revealed: The Secret Gear Connecting Google's Online Empire". Wired. June 2015.

Metcalfe's Law Revisited

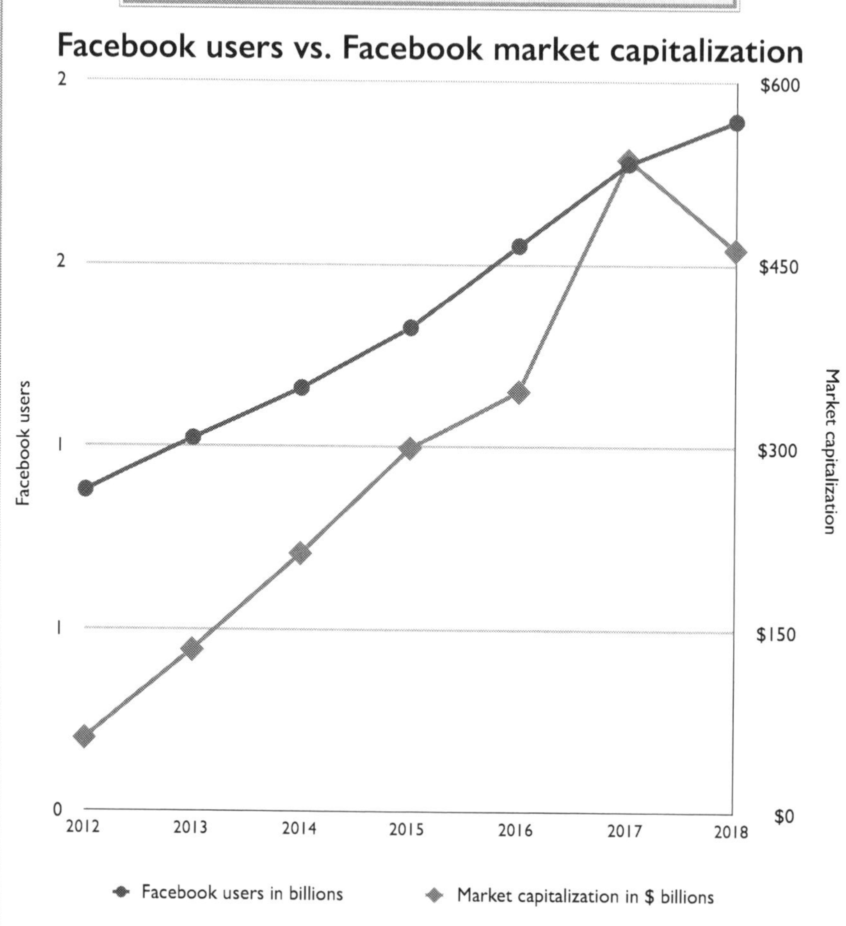

Decades after Metcalfe's Law was coined, the market success of Facebook (and other social media platforms) provide empirical proof for the theory. The growth in the number of Facebook users has grown dramatically—but the company's market cap has grown even more dramatically. (Metcalfe's Law alone isn't enough to guarantee success; the company's troubles and strife in 2018 significantly offset the positive effects of the continued growth in their user base)

CONCLUSIONS

Chapter 21: So, WTF Really Happened?

"Sometimes you will never know the value of a moment, until it becomes a memory."

Dr. Seuss

Perhaps you picked up this book before settling into seat 22A on your flight from Boston to San Francisco or Seattle to San Jose. Whether you're an MBA student, someone braving a tech startup, a 3Com alum, an alum from a rival company, an aspiring entrepreneur, or a curious onlooker, you have dedicated this time to understanding what happened at 3Com, to absorb the lessons it offers, and determine how they may apply to your own adventures.

Lessons are usually learned through the teachable moments that arise from mishaps and catastrophes. And while 3Com certainly had its share of teachable moments, the 3Com alums interviewed overwhelmingly cherish their time with the company to this day. Employees from every phase of 3Com's existence said that working for 3Com was one of the best experiences of their lives. When asked several years ago if the universe needed a book on 3Com's history, the 3Com Alumni Group on LinkedIn responded with a resounding "yes!" and the next words they offered were ones like "awesome, fun, positive," and "best experience."

The graduates of "3Com University" did indeed learn valuable lessons; the company prepared everyone for success in their next endeavors. And, as a side note, many former 3Com executives went on to become amazingly successful in their next role, with an astonishing number becoming CEOs and senior executives. (Visit 3comstory.com/interviewees for biographies of the interviewees.)

And at the outset, the book outlined some of 3Com's most important contributions to society at large. The company championed the early networks of computers of the day, creating great value and advantage to customers globally, connecting more people to more people for decades. 3Com supported invention via partnering with universities, supporting our K–12 schools, and embodying ethical principles that carried the company for 30 years.

And yes, 3Com's history is packed with its share of turmoil. In retrospect, the company that created the ubiquitous Ethernet highway for the world seemed like a magnet for near-death experiences and pivots. After boom-and-bust cycles in the 1980s, the 3Com of the 1990s grew its annual revenues from $400 million to almost $6 billion—only to plummet again, ultimately building itself back up enough to exit, thanks to its China strategy, in a respectable sale to Hewlett-Packard in 2009.

Only a small percentage of companies remain in the Fortune 500 over many decades (for proof search Wikipedia for "defunct computer companies"—Compaq, Honeywell, Netscape, Novell, Burroughs, Sun, Apollo, Silicon Graphics,

DEC, Data General, Wang, Prime ...). The constant turnover is a positive sign of dynamism and innovation, and yes, creative destruction. From winding its way around an attempt to become a computer company, to recovering its footing, to lead again in the networking industry, yet falling behind Cisco as that company's dominance grew ever stronger, 3Com's path was full of twists, turns, and missteps. Eventually, 3Com had no choice but to get out of the way.

The early growth was driven by market demand matched with fresh technological invention, led by larger-than-life leaders like Bob Metcalfe, Howard Charney, and Bill Krause who established ethical, structural, and operational habits for excellence that remained intact for much of the company's life. 3Com, as a commercial force embodying and promoting Metcalfe's Law became one of the most important startups in technology history, driving the standardization and acceptance of Ethernet via a broad range of products. 3Com didn't just develop and commercialize the new networking technology and products, it also developed new methods of distribution, innovative world-class manufacturing processes, advances in product speeds and bandwidth, and reductions in cost—all of which contributed to the Ethernet open standard achieving worldwide adoption.

3Com's early years, marked by the passion and vision of its founders, instilled in the company the value of open standards, and cooperative competition. While some offers were not meant to be, such as merging with Convergent, or Apple, others appeared to be the right course. The merger with Bridge promised a vision of networking systems products that was not exploited and resulted in great loss of talent as disgruntled Bridge employees fled to Cisco, ultimately helping that company seize the lead in the network router market, the epicenter of the network effect.

As Bill Krause's ill-fated plan to partner with Microsoft unwound along with leaving client-server computing, new CEO Eric Benhamou embarked 3Com on a new vision for global data networking. NetBuilder II, boundary routing, and stackable hubs and switches drove a successful product disruption event that even caught Cisco off guard. But Cisco's router incumbency advantage with CIOs and enterprise accounts, followed by its counter punch with its three switching players (Kalpana, Crescendo, Grand Junction), left 3Com a distant second. Ironically, former 3Com founder Charney and other 3Comers that moved to Grand Junction in the early 1990s helped drive the knockout punch.

Broadly, 3Com felt it could have success with the middle layer of the network cake, in the small and medium-sized businesses, having conceded early on the low-end hub market to SynOptics and Cabletron, and the high-end, big chassis router products to Cisco and Wellfleet.

Throughout the ups and downs, 3Com's adapter business provided the engine that enabled the company to invest in new markets like the enterprise systems business. And while adapters provided a steady stream of revenues and profits, the adapter products were hardly static; a pattern of continual improvements—like Paul Sherer's creation of the Etherlink III—successfully blocked Intel's attempts to commoditize (and then monopolize) the technology for at least the 1990s.

Eric's string of smaller, global networking systems-related acquisitions brought a series of vital technologies into 3Com, and supported its "land and expand" strategy in the UK, Israel, and the East Coast of the U.S. But 3Com's enterprise-oriented systems business continued to struggle against Cisco's first-mover advantage in routing and Fast Ethernet switching, its clever marketing, and massive sales machine. 3Com's federalist approach ultimately was less effective in the marketplace, while Cisco's model focused on executing on one segment at a time, dominating, and then moving systematically to adjacent segments. Later on, 3Com's product execution with CoreBuilder was too little, too late, and no match for Cisco's earlier success in cementing the enterprise accounts.[132]

Caught between Intel's "mobo" threat of absorbing the adapter into the PC motherboard on one side and pressure from Cisco's StrataCom purchase in the carrier market on the other, 3Com ended up acquiring USR for $6.6 billion, a large but soon-to-be-obsolete modem company; this turned out to be a toxic house of cards, triggering accounting restatements, and lawsuits. This mammoth merger of equals—driven by the board and investors anxious to see growth and profits—led Eric and 3Com into more difficult future decisions.

The Palm saga became a bit of a sideshow, albeit a profitable one. USR had acquired Palm for virtually nothing and 3Com spun it off for big billions. This created happy shareholders for a moment in time, but also resulted in another brain drain of employees leaving the mother ship. And it wasn't long before Palm suffered its own crash back down to earth.

In the unwinding of this Shakespearian tragedy, an exhausted Eric drifted away to become a successful venture capitalist, recruiting top talent from his 3Com days like Janice Roberts, Marina Levinson, and Anik Bose. The whole executive team that had "crossed the desert" with Eric, became disenchanted either with USR or with new management and exited: Chris Paisley and Bob Finocchio joined the faculty of Santa Clara University and currently serve on multiple public and private boards; Debra Engel became a guru to

132. In John Chambers' book *Connecting the Dots: Lessons for Leadership in a Startup World* (Hachette Books. 2018), he aptly states "That ability to reinvent not only your company but yourself is the most critical skill for every leader in the digital age…. Cisco's four key strengths [are]: an ability to anticipate and get ahead of market transitions, innovation processes that could be replicated at scale, a strong culture that was focused on customers, and a network architecture that gave us incredible flexibility to innovate and move into new markets."

non-profits; Doug Spreng, Alan Kessler, Ron Sege, Jeff Thermond, Paul Sherer, Rich Redelfs, Tom Werner, and others went on to run successful companies.

By the 2000s, Bruce Claflin and his new executive team were left holding the bag and set their own course to sail. That proved to be a team that valued protecting the cash on the balance sheet over product innovation, soldiering on during a disastrous recession, 9/11, and the burst of the dot-com bubble. Culture, vision, talent and trust waned as old executives bailed ship and new execs turned over. Meanwhile Broadcom and Intel absorbed 3Com's present and future adapter chip business via OEMs wishing to save a penny and Intel pushed the technology into the motherboard chipset.

Thankfully, Anik Bose and Bruce focused on H3C in China starting in 2003; that venture would ultimately become 3Com's lasting (or last, as an independent company) legacy, and an important lesson in globalization, one of the first successful joint ventures with a major Chinese tech company. The products that emerged from the H3C venture attracted HP's attention. And 3Com would ultimately be acquired by the company that, in many ways, served as its model during its startup phase.

Eventually, the radical and new networking ideas of the 1980s and 1990s became commoditized. Today's mobile and WiFi networks are taken for granted. There's little excitement about the pipes and plumbing that connect billions of computers, devices, and people to each other. The actual box that connects your home to everything else on the planet is typically a freebie provided by your service vendor. What people get excited by now are the apps on their smartphone or the media they can stream to their TV—not the networking infrastructure. But companies like the Frightful Five—Amazon, Apple, Facebook, Microsoft, and Alphabet, Google's parent, only exist because of Metcalfe's Law and Ethernet—and because of 3Com's accomplishments in making networking ubiquitous.

Chapter 22: Insights

Some of the insights from this history are specific to the company, the products, the markets, the people, and the times. But there are some key takeaways that apply broadly for all entrepreneurs, business leaders, students, and tech enthusiasts. Here are six that especially resonate:

- Passion transforms challenges into opportunities. 3Com was born out of passion, and passion, when combined with a clear vision and strategy, can be transformative. "Efforts and courage," as John F. Kennedy once said, "are not enough without purpose and direction."

- Sticky products, combined with great product execution, are worth the investment and can create an incumbency advantage. 3Com's early products were exceptionally sticky based on skilled execution, great technology, and lower cost, predisposing their customers to ignore other competitors and trump the competition.
- In fast-changing markets, you need to take risks, and anticipate the innovator's dilemma. Stated another way—the one thing that makes you successful initially, may be the very thing that makes you unsuccessful in the next round. 3Com bet the company several times. Its leaders and its employees should be applauded for decades of taking risks.
- Acquisitions work best when driven by the chosen strategy and product innovation, not personality, peer pressure, size, stock multiples, or profits. Firms can focus their strategy on being a cost leader within the targeted markets (manufacture in-house or out? Sell direct or through channels? Drive lower product costs?). They can focus on differentiation (patents, innovation, specific technologies, clever marketing). It's hard to do both really well.
- Diverse teams lead to robust insights and discovery of multiple paths to deal with problems and crises. Executive leadership must fight to keep the dialogue and communication robust and open. Culture counts.
- And finally, successful innovation sometimes requires a lot of patience. Successful small solutions may grow up to be dragons.

3Com's leaders and employees came from a diverse landscape inside and outside the U.S. Their spark and creativity contributed to 3Com's thirty-year history and to its spawning of great technological innovation. 3Comers held shared values about building important and innovative networking products that changed society for the better, while creating a sense of community and espirit de corps.

And while those who built and sustained 3Com no longer go to a building with a 3Com logo on it, 3Com's thousands of alums have gone on to be valuable contributors and trailblazers at other companies. Together, they make up an ocean of talent and experience, and continue to pioneer the next waves of technology in Silicon Valley and beyond.

Epilogue

While I was toiling away at 3Com in the 1990s, my co-author Jon Zilber was providing editorial leadership at tech magazines like *MacUser* and *The Net*. He has given back via helping the Sierra Club with its communications, and later launched Palm's social media presence. This book was only possible with the assistance of Jon's hard work, steady hand, and wise guidance. I'm immensely appreciative of all the work he has put into this.

Today I'm a retired startup CFO, enjoying book writing, family, the outdoors, and staying healthy. My journey to 3Com began first, attending USC, majoring in Accounting and Finance, and gaining my BS, MBA, and a CPA certificate. I worked for KPMG Tax in Honolulu in 1979, and then landed at Deloitte Small Business Audit Services in San Francisco in 1981. I left Deloitte in search of getting my feet wet in Silicon Valley, landing at ROLM in 1983. I was twenty-five years old, had a one-year-old son, Stephen, and my wife, Debbie, to take care of, and I joined a hard-driving internal audit practice at ROLM in Santa Clara, located at the Old Ironsides Drive campus off Great America Parkway.

At ROLM, I worked with Bob Finocchio, Ron Sege, Abe Darwish, and other ROLMans who would go on to become 3Comers. I was there from 1983 until 1987. I worked for Dan Scupin, a tough as nails former E&Y audit manager that reported to Bob Dahl, ROLM's CFO. I had interviewed with Dysan, Tandem Computer, Schlumberger, and many other Silicon Valley legends. I was thrilled to have found my way to ROLM, although it was shortly sold off to IBM for $1.25 billion in the fall of 1984, the year my daughter Caroline was born.

I had no idea that many of us from ROLM would end up being part of 3Com's history working a few blocks away off Great America in Santa Clara, as 3Com assembled a maturing enterprise management team during its explosive growth in the 1980s and 1990s.

In May of 1989, I joined 3Com as their first Internal Audit Manager, later moved into a Treasury management role, and stayed until January of 1998. I believe my audit training from ROLM not only helped me at 3Com but has also helped me with the process of piecing together this book. The methodology of forensic auditing—identifying the problem, gathering evidence, identifying causes, assessing the impact, and offering recommendations—came in handy when talking to various executives with varying views of 3Com. The opportunity to have travelled the world and examine 3Com's operations also triggered some of my own memories and insights from my rusty memory bank. And perhaps most important, the journey at 3Com provided me with rich and long-lasting friendships. The leaders and employees I spoke with, felt like we were picking up where we left off from 20 years ago or more.

Bob Finocchio, who knew my work at ROLM, was a reference for my internal audit role at 3Com in 1988, and I went to work for Judy Bruner and Chris Paisley. I thank Bob for his assistance, then and now, and for his continuing friendship over the past 35 years. Judy and Chris were wonderful mentors and friends as we learned and grew during the boom years of the 1990s and I'm grateful for working under Eric, Chris, and Judy, with their calm, intelligent force.

After the USR purchase in 1997, it felt like a slow-motion collision with an iceberg, something that would distract us, drag us down. The mood around this time was gloomy. Eric and Chris looked like they were on a death march as they had to respond to SEC and lawsuit inquiries, yet the day-to-day didn't feel like we were about to crash. Intel's Andy Groves had a mantra that seemed to describe the state of 3Com at that point: "Success breeds complacency. Complacency breeds failure. Only the paranoid survive." If there was paranoia, it was driven by the wrong reasons.

I also was interested in finding the "next wave" after 3Com. I had gotten off the finance career ladder and felt it was time to get back on with a smaller company. I was fortunate to find a fiber optic telecommunications components player called ETEK Dynamics, which was on the verge of going public. Their CFO Sanjay Subhedar took a chance on me. ETEK was later bought by JDS Uniphase Corporation, a leader in the build-out of the fiber-optic information superhighway. This was true financial serendipity, for a time. Having read George Gilder's writings about infinite bandwidth and the telecosm, I was thrilled to move on to the company that he called "the Intel of the Telecosm," notwithstanding the painful crash in 2001.

A note about Eric. While working in the 1980s, I had been exposed to mostly strong, charismatic, extrovert leaders that were only too happy to demonstrate their hard edge in leading. But Eric approached his role as though we were all faced with a sort of self-confrontation. Quietly applying his intellect, he would dissect my audit report and ask a question about an audit finding in multiple ways. This struck me as almost existential—why did this issue exist? What are the myriad different outcomes that the issue may cause, or that may have caused it? How seriously should he take it? It was a creative, building, discovering approach. This conflicted sharply with other leaders, leaders that worked more from an outer, more majestic personality, following the strict laws of economics, inputs equal outputs, always.

For the rest of my career, I had to find both paths—cultivating my inner strengths in the face of adversity, but also confronting my weaknesses, and asking the tough questions. Sometimes the Eric types don't even get the opportunity for leadership; they are often too calm, settled, and rooted in their beliefs to garner great attention. But thankfully, such leaders do still exist, and I am thrilled and appreciative to have had the chance to work with one.[133]

I am also grateful to have helped start 3Com's first United Way campaign, later expanded into a broader charitable program, with matching by 3Com. I appreciated Eric and Debra Engel's backing and found this to be a source of great personal satisfaction.

As I look back, I feel grateful that I was able to witness some of 3Com's best years in my opinion, its second decade. That decade combined the fun of working with so many great leaders and employees, with a myriad of interesting challenges and opportunities. While it was an exciting time of wealth creation for many employees and shareholders, we were all excited and energized to be participating in an industry that was changing the world in a positive way.

3Com touched greatness, as have many other Silicon Valley companies, and with some different moves earlier, might have found a different and happier outcome. But those that served 3Com all over the globe worked tirelessly, and made the decisions they felt were necessary. The leaders and all of us employees—did their level best, took some hits along the way, and have grown as well from the experience. While some may carry some anger, disillusionment, or sadness, wishing for a different outcome to the 3Com story, they still express great appreciation of their 3Com experience, and are passionate about the company to this day, proud of our accomplishments. I hope this telling of the company's story brings an understanding to what happened and why.

In addition to Jon, I wish to recognize and thank Taylor Hoefler, a University of Chicago graduate, editor, and marketing specialist, and Rose Zilber, a talented graphic design and web designer for their long hours and great help. Dan Robertson, Gerald Petak, Cindy Hawkins, Doug Spreng, Leon Woo, Jeff Thermond, the seven CEO's in this book, and countless others offered exceptional support. And Melissa Searle crafted a wonderful 3Com tennis sketch, a fun portrayal of many of the players on 3Com's court. (The inside story on that is Bob Metcalfe and Bill Krause were both top-rated players, and played frequently in the early years of 3Com.)

Thanks also to Michelle Smith, Carolyn Young, Jon Howell, Michael Swaine, Nancy Groth, and Gregg Sorensen for their help in reviewing the manuscript,

133. These comments are inspired in part by *The Road to Character*, by David Brooks, Random House, New York, 2015. David points out the differences between what he calls Adam 1 and Adam 2 personalities. Adam 1 wants to conquer, Adam 2 wants to serve the world. David's ideas, in turn, were inspired by the book *Lonely Man of Faith*, by Joseph B. Soloveitchik, Doubleday, 2006.

and to designer Kento Ikeda for transforming our manuscript into the actual book you're holding in your hands (or e-reader).

The many people who took the time to be interviewed provided fantastic and candid insights that helped me understand the real opportunities and problems that 3Com faced (you can find out more about them at 3comstory.com/interviewees): Irfan Ali, Eric Benhamou, Larry Birenbaum, Larry Blair, David Boggs, Bruce Borden, Anik Bose, John Boyle, Judy Bruner, Richard Bush, Kevin Canty, Howard Charney, Bruce Claflin, Casey Cowell, Tadhg Creedon, Abe Darwish, Tam Dell'Oro, Les Denend, David DePuy, Ajay Diwan, Michael Dolbec, Donna Dubinsky, Jerry Dusa, Debra Engel, Barry Eggers, Judy Estrin, George Everhart, Bob Finocchio, Steve Foster, Saar Gillai, Ralph Godfrey, Andrew Gottlieb, Mary Henry, Matthew Kapp, Andrew Katcher, David Katz, Alan Kessler, Ali Khan, William Krause, Cindy Hawkins, Joe Lassiter, Marina Levinson, Brad Mandell, Edgar Masri, Gwen McDonald, Bill Messer, Bob Metcalfe, Mark Michael, Cate Muether, Eileen Nelson, Chris Paisley, Gerald Petak, Sudhakar Ramakrishna, Rich Redelfs, Eddie Reynolds, Janice Roberts, Dan Robertson, Paul Rudnick, Peter Ruzicka, Don Seferovich, Ron Sege, Rich Seifert, Greg Shaw, Michael Smith, Doug Spreng, Jeff Thermond, Colleen Thuener, Bruce Tolley, David Tolwinski, Andrew Verhalen, Peter Wang, Tom Werner, and Leon Woo.

My life has been enriched and enlightened by my wife Debora, a 3Comer from 1996–2001, who assisted me in the editing and publishing aspects. Debbie and I raised our two children Steve and Caroline, along with dozens of animals over the years—horses, dogs, cats, rats, goats, and more. My family is a true dream come true—their sharing their fortitude, loving kindness, courage, creativity, and honesty, and I thank them for their patience and prayers for me throughout this process.

I thank all of 3Com's former leaders, employees, and friends of 3Com for their contributions. And I thank you, the readers, and hope that you take away some stories within it to smile about, and be glad that it happened, I hope you enjoyed this roller coaster of a ride. Happy 40th, 3Com.

Appendix A: "Does Anyone Have a Toothbrush?"

Bob Metcalfe's Eulogy of Ron Crane
Sunday, August 27, 2017

Thank you for letting me join you all in remembering and celebrating Ron, and excuse me if I cry now and then. I thought a good thing to do would be to tell a few stories that give testimony to Ron's enthusiasm, his curiosity, his intelligence, gentleness.

Let me begin with the day I first met Ron. I don't actually remember the day, but I read about it on the web, which is ironic because it was in 1975 and the web wasn't invented until 1989, but it managed to backdate it somehow... I gave a talk at Stanford and Ron came, this grad student, the story goes, and I saw Stanford as a treasure trove of people. Dave Boggs had come from Stanford and Vince Cerf was the professor that would later be known as the father of the Internet—he was the supervisor of this event. And he must have said something nice about Ron because we all came out to join him at Xerox as soon as possible.

And Ron worked on what was called the Xerox Wire, which was the renaming of the Ethernet, a rebranding of Ethernet, which we later had to undo because DEC and Intel would be damned if they were gonna sell the Xerox Wire.

In 1979, I asked three people to join me for dinner at the Bella Vista Restaurant on Skyline, which recently closed, I understand. And so it was night time, and Silicon Valley was tinkling in the background, and I proposed to John Shoch and Greg Shaw and Ron Crane that we start a company together, and we wrote it down on a napkin, and John Shoch claims to have this napkin. It's a little odd though that the napkin doesn't have a product vision on it, like it was supposed to, but it had an initial cap table in which I thought I was entitled to 51,000 shares and John Shoch, I forget the numbers, but say 18,000, and we quibbled over the numbers. John Shoch said he would not join the company, he would instead become the assistant to the president of Xerox. Greg Shaw joined immediately, and Ron said he would love to join the company, but listen to this—"but I have projects to finish at Xerox, and it'll take me about a year."

So there you see something. Commitment. His commitment to his projects. And he was not an opportunist, assuming that he viewed 3Com as an opportunity at that time. He did join the company within the year and I have to brag, the company eventually became worth, at least for a flashing moment,

$34 billion. In today's dollars, that'd be $50 billion, just for the record. And Ron was of course a founder.

In 1980 we decided, all of us, to make ethernet a standard and formed an organization called IEEE 802. That was before the dots, IEEE 802, was Project 802. And it met for the first time in February of 1980 in the Jack Tar Hotel, which isn't there anymore, it's been torn down. And Ron and I went to the 802 meetings. In fact, our company was so small, that we couldn't afford to go to all of them, so we took turns. He would go to one meeting and I would go to the next.

And then it got very political. IBM showed up, General Motors showed up, a lot of other crazy people showed up, and they started voting. It became political. And in one meeting, I was informed that I could not vote on a particular matter because I had not attended a previous meeting. They made a rule that you had to attend the previous meeting to vote in this meeting, which meant Ron alternating wouldn't fly so well. Ron and I decided he would represent 3Com at 802 for two reasons; one is one of the issues was signaling electronics and electrons and stuff like that which Ron knew much more about, but two, Ron didn't piss people off.

So he became 3Com's leader at IEEE 802 and by December of '82 we had a standard out informally at 802. And then comes 1982 and the Etherlink. We had been developing a chip with a company called Seeq called the Seeq Chip, and miraculously they delivered it on time and it worked, and we wondered what to do with it. So we asked Apple what to do with it, and they proposed that we develop an Apple Ethernet interface.

Ron wouldn't stay in the same room to discuss this topic. But we had people with our customers who believed that the customer was always right. And Apple wanted this stupid thing, and we built it, and we shipped 300 of them.

DEC on the other hand, seeing the same chips, said they would consider buying our product, and we spent the next two years teaching our attorney (Howard Charney) how to make things. But they never bought anything from us.

IBM did us the favor of saying no. No way. Which opened up a great opportunity for us, except the IBM personal computer was not in our plans, and we were very planned with Bill Krause and Larry Birenbaum and others. We were very "planful," and there was no IBM PC in our plans, and then yet here there was this thing.

So I bought a new IBM PC, found a folding table, and put it outside of Ron's cubicle. Within hours, it was completely disassembled, and we started having engineering meetings and we refined our entire engineering program around this new plan, which was to do Ethernet for the IBM/Intel/Microsoft personal computer. Just because this equipment, this curiosity that we've mentioned,

I knew I was doomed by putting this box right outside of Ron's office. It was a drug.

So the Etherlink began the two tricks there for the Seeq Chip, which allowed it to fit. The other one was the idea of putting the transceiver, which Ron was the whole expert on, onto the same board. So Ron became an engineering leader in the development of that product and as you just heard, he was sitting ... this was a cubicle on Shorebird Way, and he was sitting over there, people could see him there, but the design was not coming out with manufacturing. And we had touted our product already, the business LAN, was gonna sell it to everybody in the world starting that fall, and Ron was sitting back there at the final release of the product and he wouldn't, well, nothing was happening.

So I was deputized by the executive committee to go see Ron and see what was up because people were reluctant to talk to Ron, and you just heard why, because you couldn't have a short conversation. So I was prepared for a long one, and I went back to see how he was doing on the Etherlink, and there he was measuring the sound reflectivity of the ceiling tiles in his cubicle.

"Oh, Ron, how's the Etherlink coming?" He said, "Look, the noise is just driving me crazy, and other people here are complaining, and I'm measuring these tiles. These are the wrong tiles." But I said, "Ron, the whole company, we're burning cash toward zero, we have compulsory commitments, and you're worried about the ceiling tiles?"

Anyway, we negotiated, and I agreed that night to turn all of the ceiling tiles in this building upside down, revealing that very sound-absorbent ugly bottom part, and to buy the correct tiles forthwith so that he seemingly agreed to return to his work on the Etherlink, but still no Etherlink. He was sitting back there in that cubicle with his scopes and probes and formulas, but it isn't yet done.

So again, I was deputized to go back and have a short conversation with Ron. And it turned out he was working on a lightning protector to the circuit to put on the part. So I said, "Ron, here's the product spec. The word lightning does not appear on this spec. Nobody wants lightning protection for this Etherlink. We've never had a customer complain. Ron, drop it! Release this thing to manufacturing."

Well, Ron would not drop it, so he continued working on the lightning circuit, but eventually, in the fullness of time, Howard got the Etherlink, and Howard would make it, and eventually his product shipped millions per month and later allowed us to go public. I said eventually because I'm leaving out a big event.

We sold a thousand of these new Etherlinks to a big bank in New York, and this bank was shrewd, so they bought a thousand from us and they bought a thousand from one of our competitors, probably Ungermann Bass, I forget. And then they installed it, and they had the network running, and what do you all think happened? Lightning struck. And our competitor's cards were all fried, and our cards sat there just humming away happy as clams. And Ron could not stop smiling, vindicated of his stubborn attachment to this lightning circuit.

Ron and I both left 3Com in about 1990 or so, and at that time I remember the many times Ron had come to my office to help with management of 3Com, and I could tell it was doomed when I watched him make the turn into the cubicle row and into my office, and we were gonna sit there for two hours and he was gonna advise me on the structure of our company and so on, and we had many such meetings. But when we left 3Com, I remember him telling me he was founding his own company, LAN Media. And I remember wishing I had said: "Wow, you can run LAN Media exactly the way you wanna run it, so congratulations on this new opportunity."

After that, my contact with Ron was rather limited. Beginning in 1996, he would come for ten days to an island in Maine and we would celebrate at Big Boys Camp, as it's called, and we missed doing this Big Boys Camp this year. And Ron always came to the camp prepared to fix something, or build something. For example, we have a solar powered outhouse now.

One year, where we also had the solar power distribution system, and some of the people who worked with him on that are here today, could charge up to 24 USB-powering things.

But the best story was the time he decided that the camp needed a totem pole. So, we found a log via connections, a one-ton log that was brought over on a lobster boat and hauled it onto the beach. And we proceeded to paint it, carved an inch of wood off it, and carved shapes, which included by the way, Maxwell's Equations. And then we painted it with beautiful bright-colored paints and it was all done, and the 20 of us are standing there proud, but totem poles generally are not oriented this way. Usually they're like this (vertical). Have I mentioned how heavy this one was? So here's the twenty of us looking at a one-ton wooden object which needed to be vertical, and we had no clue about how we were gonna fix it, but our local lobsterman had a winch, and that was gonna do it. So he brought the winch, and a generator to generate electricity for the electric winch, and plugged it all in, hooked it up, all the ropes on this totem pole, and it just had to go up like this, and it didn't turn on.

So there's 20 of us standing around and the winch is not working and the totem pole was there, and Ron steps forward, and he goes over to the winch and he sits down by it, and there's all 20 of us, and we're looking like this and there it is sitting alone, and he starts taking it apart, with screwdrivers and wrenches, and the parts multiply, and there's like 50 or 100 parts and he's sitting there carefully and we're all being quiet 'cause we didn't want to disturb what was going on.

And then he takes the last parts apart and he says "Does anyone have a toothbrush?" Well, of course we ran over to the dock kits in the toilet area and we got a toothbrush and we gave it to Ron, and he went like this for a while, and then he proceeded to start assembling this winch and he got it all back together again, and he turned it on, and the totem pole went to the vertical, which it is to this very day. Does anyone have a toothbrush…?

Our children, Max and Julia, fell under Ron's influence. And what they remember most is what he taught them: he built with them a flashlight with a light and battery and a paperclip, and then he taught them by building it with them, an electric motor. When we asked our children about Ron, that's what they remember. They built a flashlight and an electric motor. So that's how Ron taught people. He would build things, and in the process of building you would learn.

So to Ron, smart, but more than smart. Curious, more than curious. Enthusiastic. Huge persistence, stamina. Two hours of a lecture on any subject on any afternoon. Quiet. Sort of the opposite of me, really. Patient. We all loved him. We're grateful that we got to know him, and may he rest in peace.

Appendix B: Timeline

Photos	Fiscal Year	Event	Dollar Amount ($)
	1964	Paul Baran invents packet switching at RAND, tested at UCLA, funded by ARPA. Used for scientists and researchers who wanted to share one another's computers remotely.	
	1968–1970	ARPANET begins operations.	
	1973	Xerox PARC Ethernet operation begins. The first Ethernet prototype is built by Bob Metcalfe, David Boggs, and Tat Lam. On May 22, 1973, Bob Metcalfe writes his memo about communicating across different ethers, including cable, telephone, and radio.	
	1974	Xerox PARC Internet Protocol operation begins. Vint Cerf and Bob Kahn publish a specification for TCP in IEEE.	
	1976	Ethernet paper published by Bob Metcalfe and David Boggs—"Ethernet: distributed Packet Switching for Local Computer Networks."	
Patent no. #4063220	1977	Ethernet Patent Issued—"Multipoint Data Communication System with Collision Detection."	
	Jun-79	Bob Metcalfe forms 3Com Corporation with help of Howard Charney, Greg Shaw, and Bruce Borden, Co-Founders. 3Com is a consulting firm for computer network technology.	
IEEE 802	Dec-79	McGraw-Hill Local Computer Network Conference, Xerox announces 860 Ethernet for 4080, IEEE forms standards committee 802.	
	1980	Xerox announces joint Ethernet project with HP, specs released, 3Com Ethernet product plan prepared, Metcalfe wins ACM Hopper Award for Ethernet. IEEE, EIS, IBM others, consider Ethernet compatibility. Bob Metcalfe gains consensus to form the critical DIX Alliance.	
	1981	3Com ships its first hardware product, the Ethernet Transceiver 3C100, later followed by the first Ethernet Adapter. Bill Krause from HP joins 3Com.	
	1984	3Com Goes Public, raises $11m, introduces 3Com network operating system.	

UNISYS	1986	3Com now selling NICs, 3Servers and its LAN software called 3+. Revenues reach $64M. Merger with Convergent Technologies fails (later bought by Unisys in 1988).	
	1987	With the help of Frank Quattrone from Morgan Stanley, who helped Bridge Communications with its first $10M financing in 1983, 3Com acquires Bridge for multiprotocol routers.	$151M
	1987	IBM's new PS/2 computers begin using Microsoft's OS2 multitasking operating system. 3Com is selling 3+ netsoftware, netstations, netadapters, and workgroup servers. 3Com signs license agreement with Microsoft, with a contract clause indicating all payments for commitments will be due upon termination of contract.	
Microsoft	1988	3Com launches 3+Open, the first network operating system based on Microsoft's LAN Manager. 3Com and Microsoft agree to co-develop and market MS OS/2 Lan Manager, and incorporate it into 3+Open.	
	Sep-90	New Renaissance Plan begins under Bill/Eric, refocus away from client-server computing, exits the networking operating systems business and 3Servers and 3Stations. CEO now Eric Benhamou, Bill Krause steps down, remains Chairman. Eric launches global data networking vision, internally and externally.	
	1992	NetBuilder II, boundary routing launches 3Com back into the large enterprise business. Key contributions by Jon Hart, Alan Kessler, Jeff Thermond, and many others.	
	1992	EtherLink III is launched, leading to great sales and profitability for 3Com.	
	Mar-92	BICC Data Networks acquisition in UK—Strengthened 3Com's structured wiring hub market, gave 3Com a hub business, brings Janice Roberts to 3Com, expanded presence in Europe.	$30M
	1993	Full scale operations commenced at its 60K sq. ft. Blanchardstown, Ireland facility.	
$	Jan-93	Star-Tek acquisition—Token Ring Hub, brings Pete Williams to 3Com, who helped lead the Stackable products success.	$48.5M
	Dec-93	High Performance Scalable Networking launched by Synernetics and Centrum Communications. Synernetics Acquisition—LANplex 6000 backbone switch, LANplex 2000 department switch, for data centers.	$104M
$	Feb-94	Centrum Acquisition—Remote access servers for Ethernet and Token Ring, sold under AccessBuilder line.	$36M

	Oct-94	Nicecom Acquisition in Israel—ATM technology. Etherlink III adapters announced.	$58.5M
	1995	Candlestick Park is renamed 3Com Park. Accessworks, and ISDN company is purchased, price not disclosed.	
$	**May-95**	Sonix Communications—UK based ISDN internetworking.	$70M
$$	**Jun-95**	Primary Access acquisition—Remote access systems provider, they sold the Aperture Platform, software defined access to public telephone networks.	$170M
$$$	**Oct-95**	Chipcom Corporation acquisition—High-end chassis hubs and switching plus IBM sales channel help.	$775M
$	**Mar-96**	Axon Networks acquisition.	$65M
$$$	**Nov-96**	OnStream Networks acquisition—ATM/broadband WAN access products.	$245M
$$$$$	**Jun-97**	US Robotics merger/acquisition—included product lines: Sportster, Courier, Palm, Megahertz, Conferencelink.	$7.3B
	Sept-97	CoreBuilder, 3Com's re-entry into layer 3 switching announced.	
	Aug-1998	Bruce Claflin named COO of 3Com, company has 13,000 employees in 182 countries.	
$$$	**Mar-1998–Dec. 1999**	Various acquisitions: • Lanwork Technologies—PC network boot technology $13M. • Euphonics—DSP-based audio technology $8.3M. • ADMtek Joint venture formed $5.3M. • Smartcode Technologies—Wireless data communications technology $17.4M. • ICS Networking (assets) Acquisition—Customer-owned tooling technology—$16.1M. • NBX Acquisition, 3Com helps evangelize VoIP—$87.8M. • LANSource Technologies acquisition—Data and Fax-over-IP software—$26M.	$173.9M
	June-00	3Com exits the large enterprise switching market, spins off Palm as a separate entity.	
$	**Mar-00**	Call Technologies acquisition—Unified messaging, operations systems and support.	$90M
	Mar-00	Palm IPO. Palm trades from IPO price of $38, to $165, before closing at $95, with market value of $53.3 billion. 3Com owns 94%, only valued at $28 billion.	
	Jul-00	Kerbango acquisition—Internet audio solution.	$80M
	2001	Bruce Claflin officially becomes CEO.	

H3C	2003	H3C Joint Venture investment—Company creates a joint venture with Huawei, or H3C, in effort to get back into large enterprise switching.	$160M
UTSTARCOM	2003	3Com sells its telecom equipment business unit CommWorks to UTStarcom for $100M, a sad ending to the USR story.	
Marlborough Campus	2003	Bruce Claflin moves 3Com Headquarters to Marlborough, MA.	
TippingPoint a division of 3Com	2004	TippingPoint acquisition—Intrusion protection security.	$430M
H3C	2006	3Com buys out Huawei's 49% stake for $882 million from the 2003 joint venture.	$882M
	Jan-2006–Aug-2006	Bruce Claflin retires, Scott Murray named CEO. Scott Murray leaves over H3C concerns, Edgar Masri named CEO.	
BainCapital	2007	Bain Capital agrees to buy the company with help from Huawei for $2.2 billion, but federal regulators block the deal.	
	Apr-08	Edgar Masri leaves, Robert Mao is named CEO and sets up shop in China where H3C claims dominance, Ron Sege named President and COO.	
	2009	3Com announces the H3C "No Compromise" Enterprise solutions globally, including the S12500 and the S9500 series.	
hp	Nov-09	3Com agrees to be acquired by HP for $2.7 billion. Deal closes on April 12, 2010.	
$$$		**Total 3Com Acquisitions**	**$10.7 Billion**

Appendix C: Required Reading

Berlin, Leslie. *Troublemakers: Silicon Valley's Coming of Age*. Simon & Schuster, 2018.

Brooks, David. *The Road to Character*. Random House, 2015.

Brunnell, David, and Adam Brate. *Making the Cisco Connection The Story behind the Real Internet Superpower*. John Wiley & Sons.

Burg, Urs von. *The Triumph of Ethernet: Technological Communities and the Battle for the LAN Standard*. Stanford University Press, 2002.

Butter, Andrea, and David Pogue. *Piloting Palm: the inside Story of Palm, Handspring, and the Birth of the Billion-Dollar Handheld Industry*. Wiley & Sons, 2003.

Chambers, John and Diane Brady. *Connecting the Dots: Lessons for Leadership in a Startup World*. Hachette Books. 2018.

Christensen, Clayton M. *The Innovator's Dilemma When New Technologies Cause Great Firms to Fail*. Harvard Business Review Press, 2016.

Estrin, Judy. *Closing the Innovation Gap: Reigniting the Spark of Creativity in a Global Economy*. McGraw-Hill, 2009.

Fisher, Adam. *Valley of Genius: An Uncensored History of Silicon Valley, as Told by the Hackers, Founders, and... Freaks Who Made It Boom*. GRAND CENTRAL PUB, 2018.

Gilder, George. *Telecosm: How Infinite Bandwidth Will Revolutionize Our World*. The Free Press, 2000.

Grove, Andrew S. *Only The Paranoid Survive*. Doubleday, 1999.

Hoefflinger, Mike. *Becoming Facebook: the 10 Challenges That Defined the Company That's Disrupting the World*. AMACOM, American Management Association, 2017.

House, Charles H., and Raymond L. Price. *The HP Phenomenon: Innovation and Business Transformation*. Stanford Business Books, 2009.

Howell, Jon P. *Snapshots of Great Leadership*. Routledge, 2013.

Karlgaard, Richard. *The Soft Edge: Where Great Companies Find Lasting Success*. Jossey-Bass, A Wiley Brand, 2014.

Maxfield, Katherine. *Starting up Silicon Valley: How ROLM Became a Cultural Icon and Fortune 500 Company*. Emerald Book Co., 2014.

Metcalfe, Robert M. *Internet Collapses and Other InfoWorld Punditry*. IDG Books Worldwide, 2000.

Moore, Geoffrey A., et al. *The Gorilla Game: Picking Winners in High Technology*. Harper Business, 1999.

Morrell, Margot, and Stephanie Capparell. *Shackleton's Way: Leadership Lessons from the Great Antarctic Explorer*. Nicholas Brealey, 2010.

Paulson, Ed. *Inside Cisco: The Real Story of Supercharged M & A Growth*. Wiley, 2001.

Smith, Randall. *The Prince of Silicon Valley: Frank Quattrone and the Dot-Com Bubble*. St. Martin's Press, 2010.

Stanley, Tim. *The Peopling of Silicon Valley, 1940 to the Present Day: An Oral History*. 2 Timothy Publishing, 2017.

Swaine, Michael, and Freiberger, *Paul. Fire in the Valley, The Birth and Death of the Personal Computer*. Third Edition. The Pragmatic Programmers, 2014.

Tinker, Bob, & Nahm, Tae Hea. *Survival to Thrival: Building the Enterprise Startup*. Vol. 1, Mascot Books, 2018.

Waters, John K. *John Chambers and the Cisco Way Navigating through Volatility*. Wiley, 2002.

Ybarra, Dano. *Guiding Your Raft*. Amazon, 2008.

Index

D

J

K

T

Y

Z

About the Authors

Jeff Chase is a Silicon Valley–based retired CFO who has worked for technology startups as well as public tech companies. Thanks to his internal audit and treasury roles at 3Com, Jeff got to know people all across the company. Their insights and recollections sparked the creation of this book. Jeff has worked for a dozen private VC-backed tech startups, as well as public companies including JDS Uniphase, 3Com, Echelon, and ROLM, along with the accounting firms Deloitte and KPMG.

Jon Zilber has been the editor-in-chief of tech magazines such as *MacUser* and *PC/Computing*. He launched and managed the social media presence for Palm, and led communications teams for nonprofits such as the Sierra Club and TechSoup. He and his teams have been recognized with Emmy, Webby, and duPont-Columbia awards. Jon is also an occasional musician, composer, and playwright.

Made in the
USA
Columbia, SC

81581857R00163